Plate Tectonics: A Very Short Introduction

VERY SHORT INTRODUCTIONS are for anyone wanting a stimulating and accessible way into a new subject. They are written by experts, and have been translated into more than 40 different languages.

The series began in 1995, and now covers a wide variety of topics in every discipline. The VSI library now contains over 350 volumes—a Very Short Introduction to everything from Psychology and Philosophy of Science to American History and Relativity—and continues to grow in every subject area.

Very Short Introductions available now:

ACCOUNTING Christopher Nobes
ADVERTISING Winston Fletcher
AFRICAN AMERICAN RELIGION Eddie S. Glaude Jr.
AFRICAN HISTORY John Parker and Richard Rathbone
AFRICAN RELIGIONS Jacob K. Olupona
AGNOSTICISM Robin Le Poidevin
ALEXANDER THE GREAT Hugh Bowden
AMERICAN HISTORY Paul S. Boyer
AMERICAN IMMIGRATION David A. Gerber
AMERICAN LEGAL HISTORY G. Edward White
AMERICAN POLITICAL HISTORY Donald Critchlow
AMERICAN POLITICAL PARTIES AND ELECTIONS L. Sandy Maisel
AMERICAN POLITICS Richard M. Valelly
THE AMERICAN PRESIDENCY Charles O. Jones
AMERICAN SLAVERY Heather Andrea Williams
THE AMERICAN WEST Stephen Aron
AMERICAN WOMEN'S HISTORY Susan Ware
ANAESTHESIA Aidan O'Donnell
ANARCHISM Colin Ward
ANCIENT ASSYRIA Karen Radner
ANCIENT EGYPT Ian Shaw
ANCIENT EGYPTIAN ART AND ARCHITECTURE Christina Riggs
ANCIENT GREECE Paul Cartledge
THE ANCIENT NEAR EAST Amanda H. Podany
ANCIENT PHILOSOPHY Julia Annas
ANCIENT WARFARE Harry Sidebottom
ANGELS David Albert Jones
ANGLICANISM Mark Chapman
THE ANGLO-SAXON AGE John Blair
THE ANIMAL KINGDOM Peter Holland
ANIMAL RIGHTS David DeGrazia
THE ANTARCTIC Klaus Dodds
ANTISEMITISM Steven Beller
ANXIETY Daniel Freeman and Jason Freeman
THE APOCRYPHAL GOSPELS Paul Foster
ARCHAEOLOGY Paul Bahn
ARCHITECTURE Andrew Ballantyne
ARISTOCRACY William Doyle
ARISTOTLE Jonathan Barnes
ART HISTORY Dana Arnold
ART THEORY Cynthia Freeland
ASTROBIOLOGY David C. Catling
ATHEISM Julian Baggini
AUGUSTINE Henry Chadwick
AUSTRALIA Kenneth Morgan
AUTISM Uta Frith
THE AVANT GARDE David Cottington
THE AZTECS Davíd Carrasco
BACTERIA Sebastian G. B. Amyes
BARTHES Jonathan Culler
THE BEATS David Sterritt
BEAUTY Roger Scruton

BESTSELLERS John Sutherland
THE BIBLE John Riches
BIBLICAL ARCHAEOLOGY
Eric H. Cline
BIOGRAPHY Hermione Lee
THE BLUES Elijah Wald
THE BOOK OF MORMON
Terryl Givens
BORDERS Alexander C. Diener and
Joshua Hagen
THE BRAIN Michael O'Shea
THE BRITISH CONSTITUTION
Martin Loughlin
THE BRITISH EMPIRE Ashley Jackson
BRITISH POLITICS Anthony Wright
BUDDHA Michael Carrithers
BUDDHISM Damien Keown
BUDDHIST ETHICS Damien Keown
CANCER Nicholas James
CAPITALISM James Fulcher
CATHOLICISM Gerald O'Collins
CAUSATION Stephen Mumford and
Rani Lill Anjum
THE CELL Terence Allen and
Graham Cowling
THE CELTS Barry Cunliffe
CHAOS Leonard Smith
CHEMISTRY Peter Atkins
CHILDREN'S LITERATURE
Kimberley Reynolds
CHILD PSYCHOLOGY Usha Goswami
CHINESE LITERATURE Sabina Knight
CHOICE THEORY Michael Allingham
CHRISTIAN ART Beth Williamson
CHRISTIAN ETHICS D. Stephen Long
CHRISTIANITY Linda Woodhead
CITIZENSHIP Richard Bellamy
CIVIL ENGINEERING
David Muir Wood
CLASSICAL LITERATURE William Allan
CLASSICAL MYTHOLOGY
Helen Morales
CLASSICS Mary Beard and
John Henderson
CLAUSEWITZ Michael Howard
CLIMATE Mark Maslin
THE COLD WAR Robert McMahon
COLONIAL AMERICA Alan Taylor
COLONIAL LATIN AMERICAN
LITERATURE Rolena Adorno
COMEDY Matthew Bevis
COMMUNISM Leslie Holmes
COMPLEXITY John H. Holland
THE COMPUTER Darrel Ince
CONFUCIANISM Daniel K. Gardner
THE CONQUISTADORS
Matthew Restall and
Felipe Fernández-Armesto
CONSCIENCE Paul Strohm
CONSCIOUSNESS Susan Blackmore
CONTEMPORARY ART
Julian Stallabrass
CONTEMPORARY FICTION
Robert Eaglestone
CONTINENTAL PHILOSOPHY
Simon Critchley
CORAL REEFS Charles Sheppard
CORPORATE SOCIAL
RESPONSIBILITY Jeremy Moon
COSMOLOGY Peter Coles
CRITICAL THEORY Stephen Eric Bronner
THE CRUSADES Christopher Tyerman
CRYPTOGRAPHY Fred Piper and
Sean Murphy
THE CULTURAL
REVOLUTION Richard Curt Kraus
DADA AND SURREALISM
David Hopkins
DANTE Peter Hainsworth and
David Robey
DARWIN Jonathan Howard
THE DEAD SEA SCROLLS
Timothy Lim
DEMOCRACY Bernard Crick
DERRIDA Simon Glendinning
DESCARTES Tom Sorell
DESERTS Nick Middleton
DESIGN John Heskett
DEVELOPMENTAL BIOLOGY
Lewis Wolpert
THE DEVIL Darren Oldridge
DIASPORA Kevin Kenny
DICTIONARIES Lynda Mugglestone
DINOSAURS David Norman
DIPLOMACY Joseph M. Siracusa
DOCUMENTARY FILM
Patricia Aufderheide
DREAMING J. Allan Hobson
DRUGS Leslie Iversen
DRUIDS Barry Cunliffe

EARLY MUSIC Thomas Forrest Kelly
THE EARTH Martin Redfern
ECONOMICS Partha Dasgupta
EDUCATION Gary Thomas
EGYPTIAN MYTH Geraldine Pinch
EIGHTEENTH-CENTURY BRITAIN Paul Langford
THE ELEMENTS Philip Ball
EMOTION Dylan Evans
EMPIRE Stephen Howe
ENGELS Terrell Carver
ENGINEERING David Blockley
ENGLISH LITERATURE Jonathan Bate
ENTREPRENEURSHIP Paul Westhead and Mike Wright
ENVIRONMENTAL ECONOMICS Stephen Smith
EPIDEMIOLOGY Rodolfo Saracci
ETHICS Simon Blackburn
ETHNOMUSICOLOGY Timothy Rice
THE ETRUSCANS Christopher Smith
THE EUROPEAN UNION John Pinder and Simon Usherwood
EVOLUTION Brian and Deborah Charlesworth
EXISTENTIALISM Thomas Flynn
EXPLORATION Stewart A. Weaver
THE EYE Michael Land
FAMILY LAW Jonathan Herring
FASCISM Kevin Passmore
FASHION Rebecca Arnold
FEMINISM Margaret Walters
FILM Michael Wood
FILM MUSIC Kathryn Kalinak
THE FIRST WORLD WAR Michael Howard
FOLK MUSIC Mark Slobin
FOOD John Krebs
FORENSIC PSYCHOLOGY David Canter
FORENSIC SCIENCE Jim Fraser
FOSSILS Keith Thomson
FOUCAULT Gary Gutting
FRACTALS Kenneth Falconer
FREE SPEECH Nigel Warburton
FREE WILL Thomas Pink
FRENCH LITERATURE John D. Lyons
THE FRENCH REVOLUTION William Doyle
FREUD Anthony Storr
FUNDAMENTALISM Malise Ruthven
GALAXIES John Gribbin
GALILEO Stillman Drake
GAME THEORY Ken Binmore
GANDHI Bhikhu Parekh
GENES Jonathan Slack
GENIUS Andrew Robinson
GEOGRAPHY John Matthews and David Herbert
GEOPOLITICS Klaus Dodds
GERMAN LITERATURE Nicholas Boyle
GERMAN PHILOSOPHY Andrew Bowie
GLOBAL CATASTROPHES Bill McGuire
GLOBAL ECONOMIC HISTORY Robert C. Allen
GLOBALIZATION Manfred Steger
GOD John Bowker
THE GOTHIC Nick Groom
GOVERNANCE Mark Bevir
THE GREAT DEPRESSION AND THE NEW DEAL Eric Rauchway
HABERMAS James Gordon Finlayson
HAPPINESS Daniel M. Haybron
HEGEL Peter Singer
HEIDEGGER Michael Inwood
HERODOTUS Jennifer T. Roberts
HIEROGLYPHS Penelope Wilson
HINDUISM Kim Knott
HISTORY John H. Arnold
THE HISTORY OF ASTRONOMY Michael Hoskin
THE HISTORY OF LIFE Michael Benton
THE HISTORY OF MATHEMATICS Jacqueline Stedall
THE HISTORY OF MEDICINE William Bynum
THE HISTORY OF TIME Leofranc Holford-Strevens
HIV/AIDS Alan Whiteside
HOBBES Richard Tuck
HORMONES Martin Luck
HUMAN ANATOMY Leslie Klenerman
HUMAN EVOLUTION Bernard Wood
HUMAN RIGHTS Andrew Clapham
HUMANISM Stephen Law
HUME A. J. Ayer
HUMOUR Noël Carroll

THE ICE AGE Jamie Woodward
IDEOLOGY Michael Freeden
INDIAN PHILOSOPHY Sue Hamilton
INFORMATION Luciano Floridi
INNOVATION Mark Dodgson and David Gann
INTELLIGENCE Ian J. Deary
INTERNATIONAL MIGRATION Khalid Koser
INTERNATIONAL RELATIONS Paul Wilkinson
INTERNATIONAL SECURITY Christopher S. Browning
IRAN Ali M. Ansari
ISLAM Malise Ruthven
ISLAMIC HISTORY Adam Silverstein
ITALIAN LITERATURE Peter Hainsworth and David Robey
JESUS Richard Bauckham
JOURNALISM Ian Hargreaves
JUDAISM Norman Solomon
JUNG Anthony Stevens
KABBALAH Joseph Dan
KAFKA Ritchie Robertson
KANT Roger Scruton
KEYNES Robert Skidelsky
KIERKEGAARD Patrick Gardiner
KNOWLEDGE Jennifer Nagel
THE KORAN Michael Cook
LANDSCAPE ARCHITECTURE Ian H. Thompson
LANDSCAPES AND GEOMORPHOLOGY Andrew Goudie and Heather Viles
LANGUAGES Stephen R. Anderson
LATE ANTIQUITY Gillian Clark
LAW Raymond Wacks
THE LAWS OF THERMODYNAMICS Peter Atkins
LEADERSHIP Keith Grint
LINCOLN Allen C. Guelzo
LINGUISTICS Peter Matthews
LITERARY THEORY Jonathan Culler
LOCKE John Dunn
LOGIC Graham Priest
LOVE Ronald de Sousa
MACHIAVELLI Quentin Skinner
MADNESS Andrew Scull
MAGIC Owen Davies
MAGNA CARTA Nicholas Vincent
MAGNETISM Stephen Blundell
MALTHUS Donald Winch
MANAGEMENT John Hendry
MAO Delia Davin
MARINE BIOLOGY Philip V. Mladenov
THE MARQUIS DE SADE John Phillips
MARTIN LUTHER Scott H. Hendrix
MARTYRDOM Jolyon Mitchell
MARX Peter Singer
MATERIALS Christopher Hall
MATHEMATICS Timothy Gowers
THE MEANING OF LIFE Terry Eagleton
MEDICAL ETHICS Tony Hope
MEDICAL LAW Charles Foster
MEDIEVAL BRITAIN John Gillingham and Ralph A. Griffiths
MEMORY Jonathan K. Foster
METAPHYSICS Stephen Mumford
MICHAEL FARADAY Frank A. J. L. James
MICROBIOLOGY Nicholas P. Money
MICROECONOMICS Avinash Dixit
THE MIDDLE AGES Miri Rubin
MINERALS David Vaughan
MODERN ART David Cottington
MODERN CHINA Rana Mitter
MODERN FRANCE Vanessa R. Schwartz
MODERN IRELAND Senia Pašeta
MODERN JAPAN Christopher Goto-Jones
MODERN LATIN AMERICAN LITERATURE Roberto González Echevarría
MODERN WAR Richard English
MODERNISM Christopher Butler
MOLECULES Philip Ball
THE MONGOLS Morris Rossabi
MORMONISM Richard Lyman Bushman
MUHAMMAD Jonathan A. C. Brown
MULTICULTURALISM Ali Rattansi
MUSIC Nicholas Cook
MYTH Robert A. Segal
THE NAPOLEONIC WARS Mike Rapport
NATIONALISM Steven Grosby
NELSON MANDELA Elleke Boehmer
NEOLIBERALISM Manfred Steger and Ravi Roy

NETWORKS Guido Caldarelli and Michele Catanzaro
THE NEW TESTAMENT Luke Timothy Johnson
THE NEW TESTAMENT AS LITERATURE Kyle Keefer
NEWTON Robert Iliffe
NIETZSCHE Michael Tanner
NINETEENTH-CENTURY BRITAIN Christopher Harvie and H. C. G. Matthew
THE NORMAN CONQUEST George Garnett
NORTH AMERICAN INDIANS Theda Perdue and Michael D. Green
NORTHERN IRELAND Marc Mulholland
NOTHING Frank Close
NUCLEAR POWER Maxwell Irvine
NUCLEAR WEAPONS Joseph M. Siracusa
NUMBERS Peter M. Higgins
NUTRITION David A. Bender
OBJECTIVITY Stephen Gaukroger
THE OLD TESTAMENT Michael D. Coogan
THE ORCHESTRA D. Kern Holoman
ORGANIZATIONS Mary Jo Hatch
PAGANISM Owen Davies
THE PALESTINIAN-ISRAELI CONFLICT Martin Bunton
PARTICLE PHYSICS Frank Close
PAUL E. P. Sanders
PEACE Oliver P. Richmond
PENTECOSTALISM William K. Kay
THE PERIODIC TABLE Eric R. Scerri
PHILOSOPHY Edward Craig
PHILOSOPHY OF LAW Raymond Wacks
PHILOSOPHY OF SCIENCE Samir Okasha
PHOTOGRAPHY Steve Edwards
PHYSICAL CHEMISTRY Peter Atkins
PLAGUE Paul Slack
PLANETS David A. Rothery
PLANTS Timothy Walker
PLATE TECTONICS Peter Molnar
PLATO Julia Annas
POLITICAL PHILOSOPHY David Miller
POLITICS Kenneth Minogue
POSTCOLONIALISM Robert Young
POSTMODERNISM Christopher Butler
POSTSTRUCTURALISM Catherine Belsey
PREHISTORY Chris Gosden
PRESOCRATIC PHILOSOPHY Catherine Osborne
PRIVACY Raymond Wacks
PROBABILITY John Haigh
PROGRESSIVISM Walter Nugent
PROTESTANTISM Mark A. Noll
PSYCHIATRY Tom Burns
PSYCHOLOGY Gillian Butler and Freda McManus
PSYCHOTHERAPY Tom Burns and Eva Burns-Lundgren
PURITANISM Francis J. Bremer
THE QUAKERS Pink Dandelion
QUANTUM THEORY John Polkinghorne
RACISM Ali Rattansi
RADIOACTIVITY Claudio Tuniz
RASTAFARI Ennis B. Edmonds
THE REAGAN REVOLUTION Gil Troy
REALITY Jan Westerhoff
THE REFORMATION Peter Marshall
RELATIVITY Russell Stannard
RELIGION IN AMERICA Timothy Beal
THE RENAISSANCE Jerry Brotton
RENAISSANCE ART Geraldine A. Johnson
REVOLUTIONS Jack A. Goldstone
RHETORIC Richard Toye
RISK Baruch Fischhoff and John Kadvany
RITUAL Barry Stephenson
RIVERS Nick Middleton
ROBOTICS Alan Winfield
ROMAN BRITAIN Peter Salway
THE ROMAN EMPIRE Christopher Kelly
THE ROMAN REPUBLIC David M. Gwynn
ROMANTICISM Michael Ferber
ROUSSEAU Robert Wokler
RUSSELL A. C. Grayling
RUSSIAN HISTORY Geoffrey Hosking
RUSSIAN LITERATURE Catriona Kelly
THE RUSSIAN REVOLUTION S. A. Smith

SCHIZOPHRENIA Chris Frith and Eve Johnstone
SCHOPENHAUER Christopher Janaway
SCIENCE AND RELIGION Thomas Dixon
SCIENCE FICTION David Seed
THE SCIENTIFIC REVOLUTION Lawrence M. Principe
SCOTLAND Rab Houston
SEXUALITY Véronique Mottier
SHAKESPEARE Germaine Greer
SIKHISM Eleanor Nesbitt
THE SILK ROAD James A. Millward
SLEEP Steven W. Lockley and Russell G. Foster
SOCIAL AND CULTURAL ANTHROPOLOGY John Monaghan and Peter Just
SOCIALISM Michael Newman
SOCIOLINGUISTICS John Edwards
SOCIOLOGY Steve Bruce
SOCRATES C. C. W. Taylor
THE SOVIET UNION Stephen Lovell
THE SPANISH CIVIL WAR Helen Graham
SPANISH LITERATURE Jo Labanyi
SPINOZA Roger Scruton
SPIRITUALITY Philip Sheldrake
SPORT Mike Cronin
STARS Andrew King
STATISTICS David J. Hand
STEM CELLS Jonathan Slack
STRUCTURAL ENGINEERING David Blockley
STUART BRITAIN John Morrill
SUPERCONDUCTIVITY Stephen Blundell
SYMMETRY Ian Stewart
TEETH Peter S. Ungar
TERRORISM Charles Townshend
THEATRE Marvin Carlson
THEOLOGY David F. Ford
THOMAS AQUINAS Fergus Kerr
THOUGHT Tim Bayne
TIBETAN BUDDHISM Matthew T. Kapstein
TOCQUEVILLE Harvey C. Mansfield
TRAGEDY Adrian Poole
THE TROJAN WAR Eric H. Cline
TRUST Katherine Hawley
THE TUDORS John Guy
TWENTIETH-CENTURY BRITAIN Kenneth O. Morgan
THE UNITED NATIONS Jussi M. Hanhimäki
THE U.S. CONGRESS Donald A. Ritchie
THE U.S. SUPREME COURT Linda Greenhouse
UTOPIANISM Lyman Tower Sargent
THE VIKINGS Julian Richards
VIRUSES Dorothy H. Crawford
WITCHCRAFT Malcolm Gaskill
WITTGENSTEIN A. C. Grayling
WORK Stephen Fineman
WORLD MUSIC Philip Bohlman
THE WORLD TRADE ORGANIZATION Amrita Narlikar
WORLD WAR II Gerhard L. Weinberg
WRITING AND SCRIPT Andrew Robinson

Available soon:

CORRUPTION Leslie Holmes
TAXATION Stephen Smith
CRIME FICTION Richard Bradford
PILGRIMAGE Ian Reader
FORESTS Jaboury Ghazoul

For more information visit our website

www.oup.com/vsi/

Peter Molnar

PLATE TECTONICS

A Very Short Introduction

OXFORD
UNIVERSITY PRESS

Great Clarendon Street, Oxford, OX2 6DP,
United Kingdom

Oxford University Press is a department of the University of Oxford.
It furthers the University's objective of excellence in research, scholarship,
and education by publishing worldwide. Oxford is a registered trade mark of
Oxford University Press in the UK and in certain other countries

First edition published in 2015

Published in the United States of America by Oxford University Press
198 Madison Avenue, New York, NY 10016, United States of America

British Library Cataloguing in Publication Data
Data available

Library of Congress Control Number: 2014957192

ISBN 978-0-19-872826-9

Printed and bound by
CPI Group (UK) Ltd, Croydon, CR0 4YY

Contents

Acknowledgements

Tanya Atwater and I began writing a book on plate tectonics in the late 1970s, but abandoned it in the early 1980s. I have incorporated a few of her turns of phrase, e.g., 'bitten by sharks' and 'dusty old wiggles'.

Many people read parts or all of the text. Latha Menon, of OUP, and Sara Neustadtl, my wife, read all chapters at least twice, John Collins, Jeff Fox, and Ellen Viste read the entire first draft, Fred Vine read the penultimate draft and corrected several errors, and Dorothy McCarthy copy-edited the final draft and caught more inconsistencies. All, especially Sara, made the book much better than it would have been without their help. Also Clarence Allen, David Battisti, Clark Burchfiel, Bob Fisher, Ann Ripley, and John Sclater read short sections and/or corrected my failing memory, and Zhang Zhuqi prepared an initial draft of Figure 36.

List of illustrations

Frontispiece: Shaded topographic map of the Earth.

Chapter 1
The basic idea

In the 1960s the Apollo missions to the Moon dominated geological, or Earth, science, but as that story was playing out in space, a bunch of mostly young, unknown scientists were leading a revolution that overturned our understanding of the Earth. Most geologists had viewed the Earth as 'solid as a rock', and therefore not deforming significantly. The surface of the Earth went up and down, somehow, and mountain ranges managed to grow in the face of erosion, which continually wore them down. Few, if any, however, realized that 20 million years from now the cities of Los Angeles and San Francisco would be suburbs of each other, or that Europe and North America would be hundreds of kilometres farther apart. Of course, exceptional geologists, while plying their various trades, advocated for large horizontal movements of continents, and now many of them rightly stand out as insightful in their recognition that 'continental drift' had occurred. Yet, for whatever reasons, 'continental drift' seems to have stirred the imaginations of relatively few geologists. Apparently most saw it as a novelty, but not a concept that would help them to solve the particular geologic problems that motivated them. The exploration of the Moon loomed as the big event in 'Earth' science. Eminent scientists boarded airplanes, then an expensive way to travel, to fly to Cape Canaveral (not yet Kennedy), Florida to watch rockets take off to the Moon.

Meanwhile, that bunch of young, unknown scientists gave us plate tectonics, and changed the way we understand the Earth. Not only do continents drift over the surface of the Earth—Europe and North America move apart at 2–3 cm/yr, while India ploughs into the rest of Eurasia at 4 cm/yr—but continental drift is but one element of a global process that includes both a steady removal of seafloor along some of its margins—punctuated by great earthquakes like those in Sumatra in 2004, Chile in 2010, and in Japan in 2011—and a continual regeneration of seafloor at 'mid-ocean ridges', like that in the middle of the Atlantic Ocean. The fundamental principle of plate tectonics is that large expanses of terrain, thousands of kilometres in lateral extent, behave as thin (~100 km in thickness) rigid layers that move with respect to each another across the surface of the Earth. The word 'plate' carries the image of a thin rigid object, and 'tectonics' is a geological term that refers to large-scale processes that alter the structure of the Earth's crust.

Many of us, when looking at a map of the world, have noticed that the east coast of South America resembles the west coast of Africa. It is hard to ignore the possibility that these two continental edges once lay against one another. Not only does the eastern corner of Brazil nestle into the bight in the West African coast near where the Niger River debouches into the sea, but convex and concave segments of the margin also mesh with one another.

The idea that the two continents were once one, or parts of one larger mega-continent, however, was not greeted with widespread enthusiasm. Although a few precocious individuals had recognized that possibility, the German meteorologist Alfred Wegener put the idea that the continents had indeed drifted apart from one another firmly into the scientific literature in 1912.

Wegener based his suggestion that the continents drifted apart on much more than just the fit of the African and South American coastlines. He noted that in fact all of the southern continents,

South America, Africa, Antarctica, and Australia, as well as the Indian subcontinent, seemed to have once formed one huge continent, which he called 'Gondwana-Land'. Geologists had found evidence of glaciation on all of these continents in rock of Permian age, approximately 300 million years old. Wegener recognized that when the continents were fitted back together, the centre of the glaciation would have lain in what is now southern Africa. He inferred that in Permian time that region lay at the South Pole. Wegener compiled other evidence that also fitted with his idea that continents had drifted, such as the distribution of plants and animals. Similarities of old fossil plants and animals could be found on various continents, but subsequently, after continents had separated from one another, those plants and animals evolved differently on the different continents. For example, it turns out that mammals, which have gained prominence since approximately 65 million years ago, did not evolve on India. We now know that India lay isolated in the middle of the Indian Ocean from approximately 120 million years ago, when together with Madagascar it separated from Africa, until approximately 50 million years ago, when it 'collided' with Eurasia. Then, shall we say, a horde of Mongolian mammals swept onto the Indian subcontinent and colonized it.

The idea that the continents once formed one, or parts of one larger, mega-continent seems to have inspired few other scientists. With hindsight, many wonder why Wegener's work was so widely ignored. Not *entirely* ignored, however; one might say that it received a hemispherically dimorphic response, with some enthusiasm in the southern hemisphere, but largely rejection in the northern hemisphere. The South African geologist, Alexander du Toit, expanded on Wegener's ideas in a book, *Our Wandering Continents*, and others closer to Gondwanaland also greeted continental drift with enthusiasm.

One obstacle to the acceptance of continental drift stemmed from an error that Wegener made. He noted that the moraines left by the last large continental ice sheets in Canada and in Scandinavia also seemed to align with one another when the margins of North America, Greenland, and Europe were fitted together. As a result, he deduced that separation of these three continental regions had begun since the last ice age, which occurred a mere 20 thousand years ago. This recent divergence of the continents we now know to be patently false; separation of Europe from North America began more than 100 million years ago. But, was he ignored because of this glaring error? Probably not.

A common view is that Wegener's difficulties in persuading others that continents had drifted stemmed from his failure to provide a mechanistic explanation for how continents actually do drift. Wegener had titled his first paper 'The Origin of Continents', and his later expanded version, a book, was entitled *The Origin of Continents and Oceans*. Although his recognition of continental drift stands as his legacy to Earth Science, he began with another profound idea: that continents and the rock beneath the oceans were different in an important way.

A decade before, the Croatian seismologist Andrija Mohorovičić had shown that the outer part of the Earth is capped by a layer that we now call the 'crust'. The light crust overlies denser mantle, which in turn overlies a denser, largely iron, core (Figure 1). Light elements like aluminum, calcium, carbon, and sodium dominate the crust, whereas the mantle not only is richer in the heavier iron and magnesium, but also contains minerals in which the elements are more closely packed than those like quartz, which are typical of the crust. Mohorovičić found that the boundary between the crust and mantle, now called the Moho in his honour, lay several tens of kilometres beneath the surface. Forty years were to elapse before marine studies could determine the thickness of the crust beneath oceans.

Wegener reasoned that beneath the oceans, the crust must be much thinner than beneath continents. To argue this, he exploited

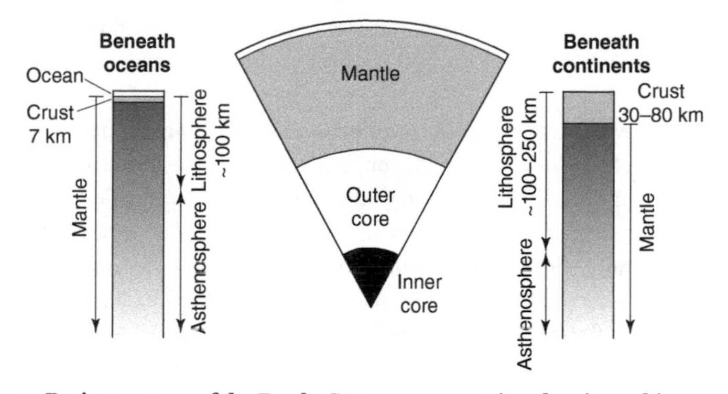

1. Basic structure of the Earth. Centre: cross section showing a thin crust (not labelled) rich in light elements, like aluminum, calcium, carbon, and sodium, the iron- and magnesium-rich mantle, and largely iron liquid outer and soft-solid inner core. Flanks: outer few hundred kilometres contrasting oceanic (left) and continental (right) regions with strong lithosphere grading downward into weaker asthenosphere.

the concept of isostasy, or Archimedes' principle applied to the Earth's crust.

Archimedes' principle manifests itself in some daily experiences. Recall that when one sees ice floating in water, 90 per cent of the ice lies below the surface of the water, whether that ice be an iceberg in the ocean or an ice cube in a glass of water. The same applies, to a first approximation, to the crust of the Earth, for which the word 'isostasy' was coined. The word 'isostasy' comes from the Greek words *isos* meaning 'equal', and *stasis* meaning 'station', but with the connotation of 'static equilibrium'. The Earth is stratified with a light crust overlying denser mantle. Just as the height of icebergs depends on the mass of ice below the surface of the ocean, so Wegener reckoned, the light crust of the Earth floats on the denser mantle, standing high where crust is thick, and lying low, deep below the ocean, where it should be thin. Wegener recognized that oceans are mostly deep, and he surmised correctly that the crust beneath oceans must be much thinner than that beneath continents. For Wegener, continents became special not just because we live on them, but also because they bore similarity to large ships at sea.

As he developed his idea of continental drift, Wegener argued that, in effect, the continents plough through the oceanic crust and mantle below. He pointed to the Andes on the west side of South America as having grown as a result of that westward push of South America on the oceanic crust to its west. As for a force to move the continents, Wegener appealed to the centrifugal force that tends to push objects away from the axis of rotation, or the poles. Wegener imagined that outward flow away from the poles tended to move continents toward the equator, though much of the movement that he imagined was not equatorward.

The rejection of continental drift in the northern hemisphere derived in part from Wegener's inability to offer a sensible mechanism for such drift, for the centrifugal force was obviously inadequate, but as well from the negative reaction of Harold Jeffreys, an outstanding mathematical physicist who repeatedly made ground-breaking contributions to Earth Science. Jeffreys's credibility rested on diverse accomplishments. He had carried out fundamental studies of fluid mechanics and of ocean circulation. An eminent statistician, he and the Australian geophysicist Keith Bullen compiled the arrival times of seismic waves recorded by seismographs despite their inaccurate timing to construct a remarkably accurate image of deep-Earth structure. Jeffreys stood virtually alone against the giants of statistics of his time in his embrace of Bayesian statistics, an idea suggested by Thomas Bayes in the 18th century that in drawing inference one could add bias, if sensible probabilities were assigned to that bias; today, most statistical analyses exploit Bayes's theorem. A polymath unmatched among contemporary Earth scientists, Jeffreys unhesitatingly expressed his opinions, which were commonly based on sound physics and good mathematics. He also believed, to his death in 1989, that rock was too strong to allow continents to drift with respect to one another.

Meanwhile, Wegener died during the winter of 1930 while carrying out meteorological and glaciological fieldwork in Greenland.

In the face of strong opposition, continental drift gained few footholds in the 1930s and 1940s, but in the 1950s, a new development, palaeomagnetism, allowed a test. When it forms, rock can become magnetized parallel to the Earth's large-scale magnetic field (Figure 2). Cooling lavas, for example, become magnetized as the temperature drops low enough, and the abundance of iron oxides in lavas allows the rock to become magnetized parallel to the Earth's magnetic field. Sedimentary rock also can become magnetized parallel to the Earth's field. That field points up at the south magnetic pole, toward the north across the equator, and then down at the north magnetic pole. The north and south magnetic poles do not lie at the axis of rotation of the Earth, the North and South Poles. The Earth's field, however, drifts slowly westward, and when averaged over thousands of years the average direction of the magnetic field is from South Pole to North Pole. From a measurement of the direction in which a hunk of rock is

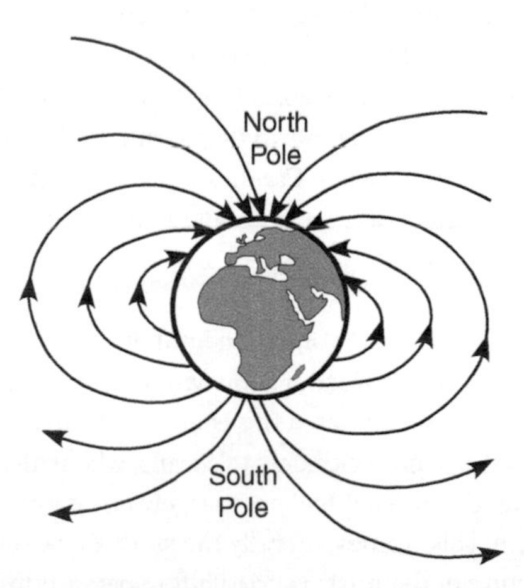

2. Earth's magnetic field, with lines of force pointing upward at the South Pole, trending horizontally and northward at the equator, and downward at the North Pole.

magnetized, one can infer where the North Pole lay relative to that rock at the time it was magnetized. It follows that if continents had drifted, rock of different ages on the continents should be magnetized in different directions, not just from each other but more importantly in directions inconsistent with the present-day magnetic field.

In the 1950s, several studies using palaeomagnetism were carried out to test whether continents had drifted, and most such tests passed. As an example, if new lava erupted onto India today, it should become magnetized pointing downward and toward the North Pole. Measurements of 65-million-year-old lava, however, showed its magnetization pointing upward and approximately northward. That rock, along with the entire Indian subcontinent, lay in the southern hemisphere when it was magnetized.

Palaeomagnetic results not only supported the idea of continental drift, but they also offered constraints on timing and rates of drift, which hitherto had been few. Yet, their impact on the Earth sciences was modest. When one of the giants of palaeomagnetism, Edward (Ted) Irving, defended his PhD thesis in Cambridge University in 1955, he failed. As he summarized the event, he failed to provide a dissertation that was acceptable to the examiners. The matter was corrected a decade later when Cambridge awarded him an honorary degree, but the implication for continental drift was obvious: its time had not yet come.

Then in the 1960s, the idea of continental drift saw a renaissance, but subsumed within a broader framework, that of plate tectonics. Three major events precipitated this change: a switch in emphasis, and relevant data, from continents to oceans, where discoveries were being made rapidly in what had been largely unexplored territory; rapid growth in seismology, literally the study of earthquakes, but also the structure of the Earth; and a shift in perspective from the chemical stratification of the Earth, in terms of crust and mantle, to another that emphasized strength—a strong lithosphere (from the

Greek word *lithos* meaning 'rock') overlying a weak asthenosphere (Greek *asthenos* meaning 'weak').

This book concerns the recognition of plate tectonics, including where it has unified disparate topics and thinking, as well as where it has failed, with separate chapters devoted to separate elements. The subject, however, is not linear, with one aspect logically following from another. So, in the rest of this chapter, I introduce some basic aspects of plate tectonics that will, I hope, make it possible to mention ideas, processes, features, and phenomena before they are discussed in more detail in subsequent chapters.

Following the Second World War, in which naval battles played a key role, industrial nations began vigorous study of the deep ocean. At first, attention was focused largely on mapping the depth of the ocean, or the shape of the ocean floor. Despite the wide spacing between the tracks that ships had taken in this new age of exploration, by the mid-1950s two features that were to play a major role in plate tectonics had been recognized: 'mid-ocean ridges' and 'fracture zones'.

Broad regions of shallow bathymetry define a globally encircling 'ridge' or 'rise' (see Frontispiece). The Mid-Atlantic Ridge lies midway between North America and Europe and between South America and Africa. This 'mid-ocean ridge' continues into the Indian Ocean, where it splits into two. One continuation trends northwestward into the Gulf of Aden and the Red Sea between Africa and Arabia, and the other southeastward between Australia and Antarctica, and then into the Pacific. In the Pacific, the 'ridge' does not lie midway between continents and is a much wider, gentler feature than in the other oceans; it is called the East Pacific Rise.

This difference between the Atlantic with its 'ridge' and Pacific with its 'rise', though now understood well and discussed in Chapter 2, retarded progress somewhat. The scientists who studied the former were based at the University of Cambridge, Lamont Geological Observatory (now Lamont-Doherty Earth Observatory) of Columbia University, and Woods Hole Oceanographic Institution in Woods Hole, Massachusetts, but those studying the Pacific sailed largely from Scripps Institution of Oceanography, now a part of the University of California in San Diego. Exploiting the analogy with blind men examining an elephant, we might say that one group studied the trunk and the other the tail.

Today the mid-ocean ridge system is so obvious on a chart showing the bathymetry of the global ocean, like that in the Frontispiece, that it is hard to imagine the imagination needed to recognize this feature when only sparsely spaced ship tracks crossed it. The key came from completely different work. The French seismologist J.-P. Rothé had noticed that a belt of earthquakes follows the axis of the Atlantic Ocean, along the crest of the Mid-Atlantic Ridge (Figure 3), and continues into the Indian Ocean and Pacific, essentially where those few sparse ships' tracks showed a shallow seafloor. Maurice Ewing and Bruce Heezen at the Lamont Geological Observatory then took the leap and postulated a continuous globe-encircling mid-ocean ridge system despite gaps in bathymetric coverage as large as 1000 km.

In addition, along much of the Mid-Atlantic Ridge and its continuation into the Indian Ocean, a narrow valley, 20 to 40 km wide, marks the axis of the ridge. When in the 1950s Ewing and Heezen recognized this axial valley, they also saw that it resembled the valley that defines the East African Rift System. There a valley has formed between two high, gentle surfaces, and is reminiscent of a keystone in an arch that has dropped down slightly as the sides of the arch moved apart. They called the axial valley of the Mid-Atlantic Ridge a 'rift' and correctly inferred that seafloor on opposite sides of the ridge had moved apart. They disagreed, however, about the significance of not only the rift, but also the

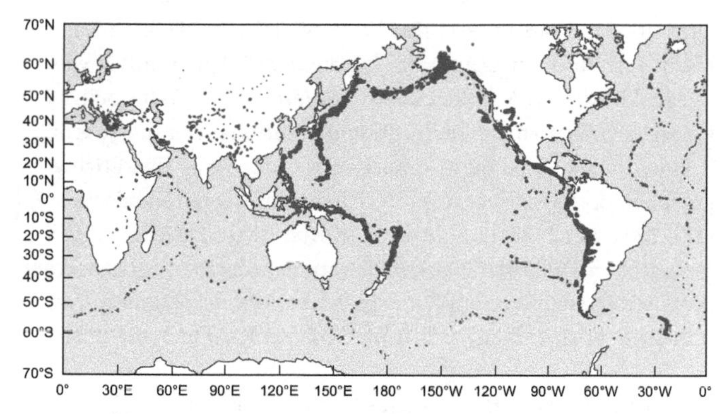

3. Map of the world's earthquakes, or seismicity. Dots mark epicentres of earthquakes. Belts of seismicity outline plates (Figure 6). Only within continents is earthquake activity widespread.

entire mid-ocean ridge system. Ewing advocated modest crustal extension, as in Africa, while Heezen argued that the mid-ocean ridge system resulted from an expanding Earth.

In his presidential address to the Geological Society of America in 1960, Harry Hess of Princeton University merged the growing facts about the oceans with the old ideas of continental drift. He proposed that new seafloor was made at mid-ocean ridges where the two flanks of the ridge diverged from one another, and where lava erupted at the surface to fill the void that would be created by the diverging seafloor. New seafloor formed continually as older seafloor moved away from the ridge axis. Continents moved apart along with that seafloor. Although a field geologist who worked largely on land, Hess was familiar with the ocean and its floor. While Commander of a transport vessel serving troops in the Pacific during the Second World War, and from 'discrete choices of travel routes—perhaps not always in strict accord with orders—and continuous use of the equipment', he took note of the bathymetry beneath his ship as it traversed the Pacific many times.

Hess titled the paper presenting his views 'History of Ocean Basins', but he introduced it as an 'essay in geopoetry'. Apparently it was rejected by the *Geological Society of America Bulletin* for being too speculative to merit publication in a serious scientific journal, despite the paper being his presidential address as the retiring president of that society in 1960. The paper was not published until 1962, in a book honouring A. F. Buddington. Meanwhile, in 1961, Robert Dietz of the US Navy National Electronics Laboratory proposed the name 'ocean floor spreading', which soon morphed into 'seafloor spreading', for the process that Hess had imagined.

Chapter 2 discusses the evidence that demonstrated seafloor spreading and that took the idea far beyond what Hess or Dietz imagined.

Also in the 1950s, H. W. (Bill) Menard, at both the National Electronics Laboratory and Scripps Institute of Oceanography, together with Dietz mapped a second feature important to the development of plate tectonics: huge linear topographic scars in the seafloor bathymetry across much of the eastern Pacific, which they dubbed 'fracture zones' (see Frontispiece and Chapter 3). The linearity of fracture zones suggested to Menard and Dietz that they marked major fractures, along which the expanse of seafloor on one side slipped horizontally past the other, what in geological terminology are called 'strike-slip faults' (Figure 4). The 'strike' of the fault is the orientation, or azimuth, of its trend, and slip on a strike-slip fault is horizontal and parallel to that trend. The San Andreas fault in California is a familiar, modern example of such a fault. Menard and Dietz's logic concerning strike-slip faulting was

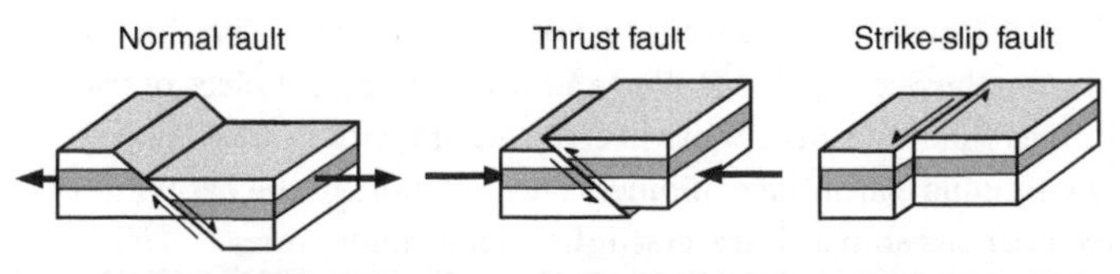

4. Block models illustrating slip on normal, thrust, and strike-slip faults.

sound, but for reasons discussed in Chapter 3, they inferred the wrong sense of motion.

Though not apparent to more than a few in 1960, the slow deformation of the Earth's crust to build mountain ranges does not occur slowly and steadily, but abruptly in earthquakes. If one wants to study deformation of the Earth's crust in action, the quick and dirty way is to study earthquakes. For example, although the seafloor east of Japan has been sliding westward beneath the island for more than 100 million years, that slip is not steady on human time scales. A portion of the Japanese island of Honshu lurched 20 (to, in one area, as much as 50) metres over the Pacific Ocean floor in the Tokohu-Oki earthquake of 2010, when a tsunami damaged the nuclear power plant at Fukushima.

Until the 1960s, studying fracture zones in action was virtually impossible. Nearly all of them lie far offshore beneath the deep ocean. Then, in response to a treaty in the early 1960s disallowing nuclear explosions in the ocean, atmosphere, or space, but permitting underground testing of them, the Department of Defense of the USA put in place the World-Wide Standardized Seismograph Network, a global network with more than 100 seismograph stations. Seismology needed modernization, and seismographs were needed to detect underground nuclear explosions, and to discriminate them from earthquakes. Suddenly remote earthquakes, not only those on fracture zones but also those elsewhere throughout the globe (Figure 3), became amenable to study. As discussed in Chapters 3, 4, and 5, the study of earthquakes played a crucial role in the recognition and acceptance of plate tectonics.

Wegener had imagined that the thick crust of continents floating on a weak substratum ploughed through thinner, weaker oceanic crust and upper mantle; in plate tectonics continents merely became passive passengers riding on strong 'plates' of lithosphere, the outer ~100 km, to as much as ~250 km, of the solid Earth

(Figure 1). We call them plates, but more precise words would be 'spherical caps' of lithosphere.

Like ice cream or butter, most solids are strongest when cold, and become weaker when warmed. Temperature increases into the Earth. As a result the strongest rock lies close to the surface, and rock weakens with depth. Moreover, olivine, the dominant mineral in the upper mantle, seems to be stronger than most crustal minerals; so, in many regions, the strongest rock is at the top of the mantle. Beneath oceans where crust is thin, ~7 km, the lithosphere is mostly mantle (Figure 1). Because temperature increases gradually with depth, the boundary between strong lithosphere and underlying weak asthenosphere is not sharp. Nevertheless, because the difference in strength is large, subdividing the outer part of the Earth into two layers facilitates an understanding of plate tectonics.

Reduced to its essence, the basic idea that we call plate tectonics is simply a description of the relative movements of separate plates of lithosphere as these plates move over the underlying weaker, hotter asthenosphere. As illustrated in cartoon-fashion in Figure 5, plates separate at mid-ocean ridges, as Hess had imagined. They slide horizontally past one another at what are called transform faults. Menard and Dietz's fracture zones, discussed in detail in Chapter 3, are scars in the seafloor created by slip on transform faults. Finally, one plate plunges beneath another at 'subduction zones' (Chapter 4), marked by deep-sea trenches, abundant earthquake activity, and volcanoes that in many cases form arcuate chains ('island arcs') like the Aleutians, or the Lesser Antilles at the eastern edge of the Caribbean Sea.

Most of the Earth's surface lies on one of the ~20 major plates, whose sizes vary from huge, like the Pacific plate, to small, like the Caribbean plate (Figure 6), or even smaller. Narrow belts of earthquakes mark the boundaries of separate plates; compare the

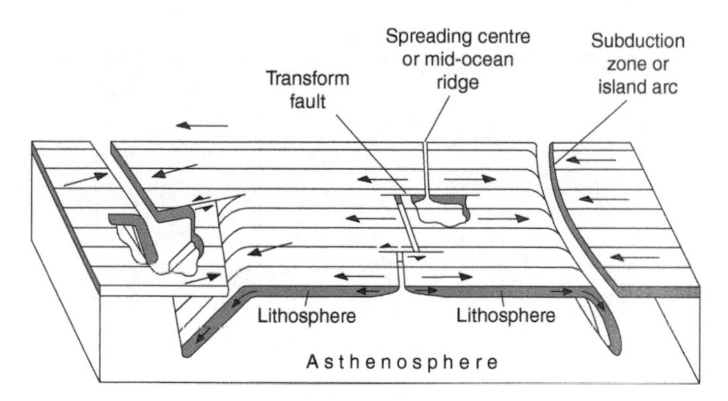

5. Block diagram showing plates of lithosphere moving with respect to one another. Two plates in the middle separate from one another at a spreading centre or mid-ocean ridge, slide past one another at transform faults, and plunge beneath plates on the right and left. These four plates resemble those in the South Pacific, where the Pacific and Nazca plates diverge (defined in Figure 6), with the former plunging beneath the Australia plate on the left and the latter beneath the South America plate on the right.

map of earthquakes (Figure 3) with that of plates (Figure 6). The key to plate tectonics lies in these plates behaving as largely rigid objects, and therefore undergoing only negligible deformation. That rigid-body movement is the subject of Chapter 5. Some plates, like the Pacific, Nazca, and Cocos plates, consist entirely of oceanic lithosphere, with virtually no continental material at all. Large continents, like Eurasia or Africa, however, occupy large fractions of the plates on which they lie. Ironically, whereas Wegener focused on continents in presenting his idea of continental drift, plate tectonics enjoys its widespread success largely in regions below deep oceans; as discussed in Chapter 6, it fails most spectacularly within some portions of continents. The northern boundaries of the Africa, Arabia, and India plates lie hundreds to thousands of kilometres south of the southern boundary of the Eurasia plate (Figure 6); between them the rules of plate tectonics do not help us much to understand how the intervening regions deform.

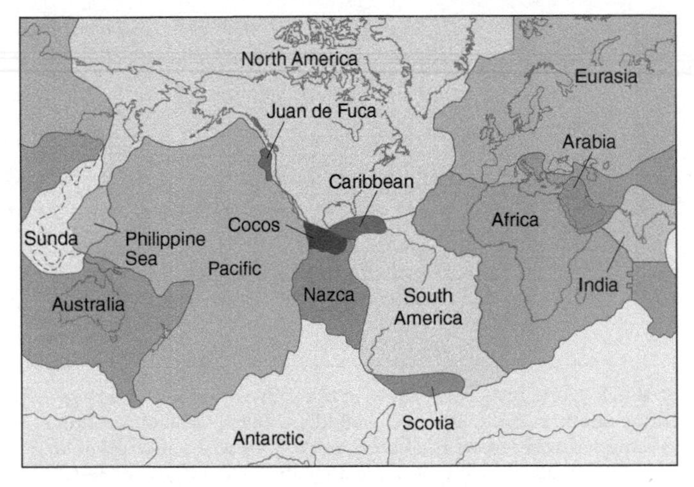

6. Map of major lithospheric plates.

By the early 1970s, the basic elements of plate tectonics had permeated essentially all of Earth science. In addition to the obvious consequences, like confirmation of continental drift, emphasis shifted from determining the history of the planet to understanding the processes that had shaped it. Accordingly, geologists who had mapped *terra incognita* as an end in itself increasingly looked to terrains that provided field laboratories to study processes. The entire globe became a target of study, and geology rapidly shifted from being a branch of 19th century natural history to becoming a 20th century quantitative physical science.

Chapter 2
Seafloor spreading and magnetic anomalies

With his paper buried in a book, Hess's ideas comprising seafloor spreading might have gone unnoticed had he not met Fred Vine, then a graduate student at Cambridge University. Hess was visiting Cambridge for a sabbatical year, and Vine, perhaps alone, constituted the receptive audience that he needed.

The Vine–Matthews Hypothesis

In 1963, Vine and his PhD advisor at Cambridge, Drummond Matthews, took seafloor spreading far beyond what Hess was likely to have imagined. They started by accepting (or assuming) Hess's suggestion that the seafloor spreads. When the mantle beneath the mid-ocean ridge melts, the molten material rises as lava, and when it reaches the surface, it freezes to become the dark-coloured rock, basalt. Following the ideas of palaeomagnetism, Vine and Matthews assumed that when the newly added lava cooled and froze to become solid, it would become magnetized in the direction of the Earth's ambient magnetic field. We now know that for the past 50 million years, the Earth's magnetic field has reversed itself randomly at an average rate of approximately three times per million years. Vine and Matthews assumed that the field had been reversed in the past, at best a contentious idea in 1963, and that such reversals from normal to reversed and back again had occurred. Immediately after the Earth's field has reversed polarity, they reasoned, new material added along the axis of the spreading centre would become magnetized in the opposite direction from that of the slightly older basalt on either side of the axis. From these assumptions—seafloor spreading, strong magnetization of basalt parallel to the Earth's ambient field, and reversals of the magnetic field—they predicted that strips of seafloor magnetized with 'normal' and 'reversed' polarity would be accreted to the edges of the plates (Figure 7), though the idea of strong plates of lithosphere had not yet been included. These strips of

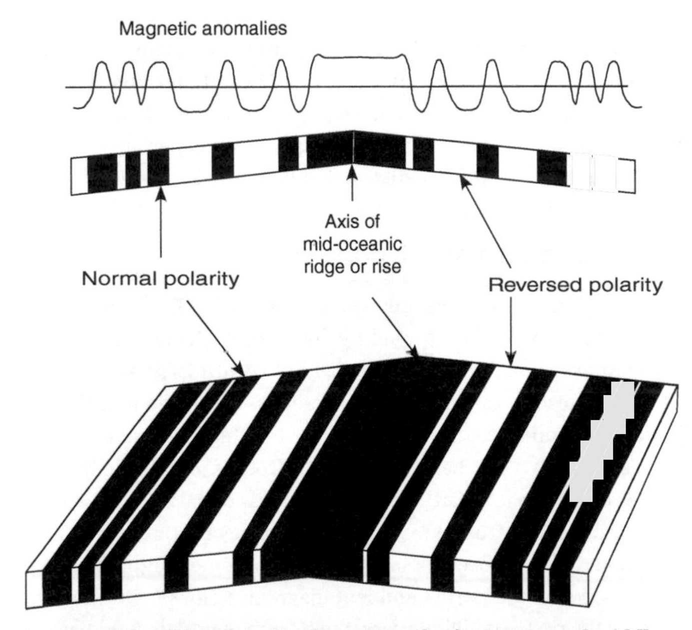

7. Essentials of the Vine–Matthews Hypothesis. Bottom and middle: block diagram and cross section, respectively, of seafloor across a mid-ocean ridge. Black shows seafloor magnetized parallel to the Earth's present-day magnetic field ('normal' polarity), and white regions where magnetized in the opposite direction ('reversed' polarity). Top: magnetic anomalies, recordings of small perturbations to the magnetic field due to the magnetized seafloor.

alternating normal and reversed magnetization perturb the strength of the Earth's magnetic field in what are called 'magnetic anomalies', tiny aberrations in the magnetic field that ships can measure as they sail across the magnetized ocean floor.

When Vine and Matthews made these arguments, strips of positive and negative magnetic anomalies had been recognized, but in the northeastern Pacific where their relationship to the mid-ocean ridge system was not obvious. In 1966, however, James Heirtzler and Xavier Le Pichon, then at Lamont Geological Observatory, and J. G. Baron of the US Naval Oceanographic Office published a map of magnetic anomalies along the Mid-Atlantic Ridge south of Iceland that showed the strips that Vine and Matthews had predicted (Figure 8).

Each of Vine and Matthews's basic assumptions is now well established, but at the time in 1963, when seafloor spreading seemed like madness, the Vine–Matthews Hypothesis must have seemed even yet more absurd. Although quantifying and testing the hypothesis required several additional assumptions and simplifications, what is most extraordinary about the hypothesis is that it not only survives these simplifications, but also allows them to be bounded quantitatively.

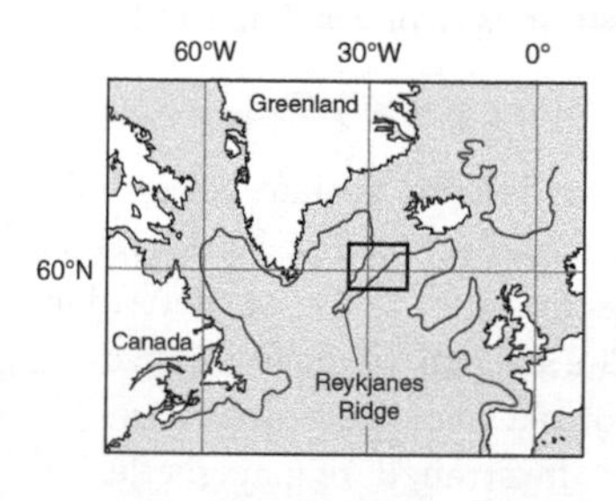

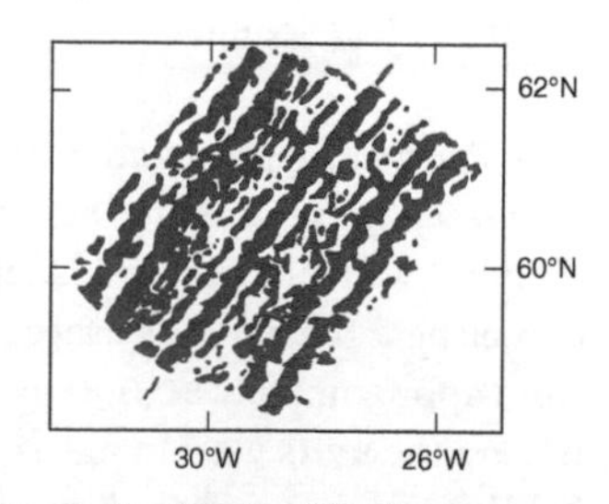

8. Linear magnetic anomalies (right) south of Iceland in the region in the box in map (left). Black shows positive anomalies, and white negative anomalies.

Predicting the shapes and distribution of magnetic anomalies using Vine and Matthews's hypothesis requires knowing the history of reversals of the Earth's field, but in 1963 a clear demonstration that the magnetic field had been reversed was lacking, to say nothing of the existence of an accurate history of reversals. Progress, however, was fast. By the mid-1960s, a fairly accurate history of reversals for the past 3 million years had been measured. By the late 1960s, when seafloor spreading and the Vine–Matthews Hypothesis had been accepted, the logic was turned around; measurements of magnetic anomalies by ships traversing the oceans allowed a reliable extrapolation that yields a history of reversals of the Earth's field over the last 150 million years. But, let's back up.

Magnetic anomalies are but minor aberrations of the Earth's magnetic field (Figure 2), which sailors have used to navigate for millennia. That magnetic field is generated deep in the Earth by fluid motion in its liquid, largely iron, core (Figure 1).

Most lava that has been erupted onto the Earth's surface, whether on land or on the ocean floor, cools quickly, in periods of minutes to maybe days. During that cooling, minerals containing iron align with the magnetic field that originates in the Earth's core. Thus, the frozen lavas become magnetized in geologically short periods, and it turns out that the magnetization is strong, as Vine and Matthews had assumed.

To a first, good approximation, we can describe the Earth's magnetic field as a large value that varies smoothly over the globe in a way that we know well, but with small deflections due to the magnetization of rock near the Earth's surface. The strength of the field ranges from approximately 50,000 nanotesla (nanotesla, abbreviated nT, are the units used to measure the strength of a magnetic field) at the poles, where it points vertically upward at the South Pole and downward at the North Pole, to 25,000 nT at the equator, where the field is horizontal and points from south to north. By comparison, the magnetization of basalt on the seafloor produces a small anomalous field of approximately 100 nT at the sea surface. Thus, near the poles, the anomalous field, or the 'magnetic anomaly', is only 0.2 per cent of the total field strength, obviously only a small perturbation to the Earth's field.

We measure the strength of the magnetic field with a magnetometer. By the 1960s, remarkably simple technology made measuring the field easy. The proton precession magnetometer is little more than a bottle of water with coils of wire wrapped around it. Nuclei of hydrogen atoms, protons, in the water (H_2O) molecules carry a tiny magnetization. In the presence of a strong magnetic field, the tiny magnetization associated with a proton becomes aligned parallel to that strong field. So, a strong electrical current is sent through the coil of wire to create a strong magnetic field in the bottle of water, in order to align the magnetizations of the protons. Then, the electrical current is abruptly shut off, so that the water molecules in the water bottle sense only the Earth's much weaker magnetic field. The Earth's field is too weak to align the magnetizations of the protons. Instead they wobble, or 'precess', around the direction of the Earth's field at a rate that is proportional to the strength of the Earth's field. The precessing protons change the magnetic field around the bottle of water, and that changing field then induces an electrical current in the same wires wrapped around the bottle. The frequency at which the induced current varies is proportional to the strength of the Earth's field. Wonderfully tough and reliable, the proton precession magnetometer can be towed behind the ship, far from sources of magnetic field on the ship, through all weather and all seas, bitten by sharks, banged and beaten, and continue through it all to measure the magnetic field several times per minute with an accuracy of one nanotesla or better.

Because measuring magnetic anomalies was easy, by the early 1960s towing a magnetometer had become a routine on oceanographic ships whenever they happened to be steaming from one place to another. Someone on board dutifully copied down the value of that strength every minute as the ship steamed along. The people collecting the data noticed the variations in the strength of the field measured by the magnetometer, but making sense of the measurements was another matter. Much of the data was just stacked in a corner, in case it might be useful in the future. Maurice Ewing, who more than anyone was responsible for the extensive amount of marine geophysical data that existed in the mid-1960s, supposedly once said of the magnetometer, 'See no evil, hear no evil, speak no evil.' Because the data were so easy to obtain, their apparently limited value ought not to prevent

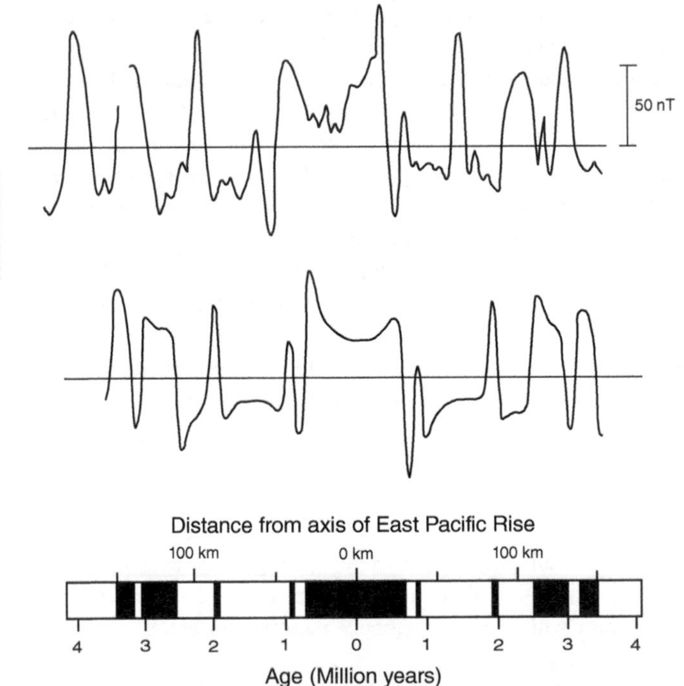

9. Observed (top) and calculated (middle) magnetic anomalies across the East Pacific Rise in the South Pacific Ocean. Calculated anomalies are based on the sequence of polarities (black is normal and white is reversed) at the bottom.

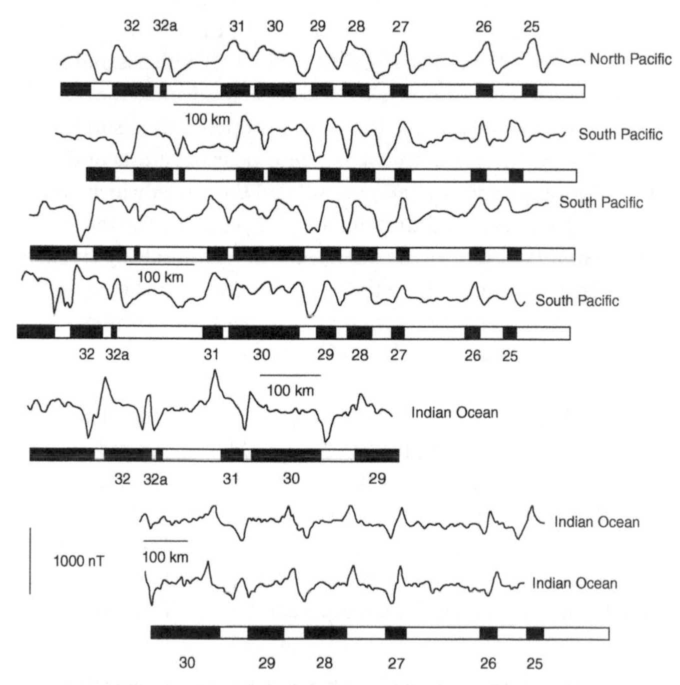

10. Magnetic anomalies and corresponding magnetization of underlying seafloor (black is normal polarity and white reversed polarity) over seafloor of the same age in the North Pacific, South Pacific, and Indian Oceans. Numbers beneath blocks of normal polarity are the 'names' given to the anomalies.

their collection. Accordingly, a great many magnetic profiles had been collected, primarily because it was easy to do (so why not?), during the decade before their meaning and importance was recognized.

Then, when marine geologists began to take the Vine–Matthews Hypothesis seriously, those dusty old measurements came alive and began to speak. Ewing's ships had been circling the oceans for many years making a spider web of ship tracks, so that the groups at the institution that he had founded and directed, the Lamont Geological Observatory, had an especially complete collection. Armed with the Vine–Matthews Hypothesis and an approximate history of reversals of the magnetic field for the past 3 million years, the marine geomagneticists were suddenly able to deduce the age of the ocean floor near the crests of the mid-ocean ridges and rises (Figure 9), and the rate at which new ocean floor is created at different spreading centres. The rate at which the seafloor is created can be determined simply by dividing the distance from the ridge axis by the age: rate = distance/age. This was a huge step for continental drift, because finally one could calculate rates at which continents moved apart.

Moreover, because magnetic anomalies over older ocean floor resemble one another (Figure 10), they could be correlated from ocean to ocean. Within a few short months in 1966–7, by extrapolating the rates at which seafloor had been created since 3 million years ago to this older seafloor, marine geologists were able to work out the basic history of large parts of the ocean floor, once they knew the ages of magnetic anomalies older than 3 million years.

Dating the ocean floor with magnetic anomalies

When Vine and Matthews proposed that strips of seafloor parallel to the mid-ocean ridges were magnetized in opposite directions, because the Earth's field had reversed itself many times, few others were convinced of such a reversing field. By 1966, however, largely through the work of Allan Cox, Brent Dalrymple, and Richard Doell, then at the US Geological Survey, and Ian MacDougall, François Chamalaun, and Donald Tarling, all then at the Australian National University in Canberra, reversals of the Earth's magnetic field had become an established fact, and a preliminary chronology of reversals for the last 3 million years had been determined. Cox, Dalrymple, Doell, MacDougall, Chamalaun, and Tarling had sampled lavas from volcanoes, and brought samples back to the laboratory to measure both the ages of the lavas and the orientation of their magnetization. The strong magnetization of the volcanic rock made it relatively easy to measure the magnetization. The breakthrough for this work was the development of accurate dating of the lavas, using the radioactive decay of potassium to argon.

This work, from studies of volcanic rock on land, not only demonstrated that the Earth's field had reversed, but also determined the history of reversals. Thus, one assumption of the Vine–Matthews Hypothesis was established. It also happens that oceanic rock, basalt, is one of the most highly magnetic rock types. Thus, when new basaltic lava pours out along a mid-ocean ridge to form new oceanic crust, it freezes rapidly, cools, and becomes strongly magnetized in the direction of the existing magnetic field. Hence, by 1966, a test of the Vine–Matthews Hypothesis became a test of Hess's seafloor spreading. In December 1966, Vine, then a young instructor at Princeton, published, apparently with help from Hess to prevent rejection of the paper, what many of us call Vine's bible of magnetic anomalies, which confirmed seafloor spreading.

The age of the ocean floor beneath the particular anomalies could be determined by correlating the sequence of magnetic anomalies with the reversal chronology measured on land (Figure 9). With the chronology of geomagnetic reversals known for the past 3 million years, an average rate at which new seafloor had been created at a particular mid-ocean ridge or rise could be obtained for this interval. Most of the ocean floor, however, is much older than 3 million years. Magnetometers towed behind ships steaming over this ocean floor measured magnetic anomalies that resembled one another, both along ships' tracks that were near each other and in different oceans (Figure 10). So, the Vine–Matthews Hypothesis seemed to apply to that older seafloor. Similar sequences of magnetic anomalies from the different oceans were presumed to overlie seafloor of similar ages. Moreover, that seafloor surely was older than 3 million years, but how old?

Estimating the age of the older seafloor led to one of Earth Science's boldest extrapolations. Jim Heirtzler, three graduate students, Walter Pitman, Geoff Dickson, Ellen Herron, and a recent recipient of his PhD, Xavier Le Pichon, extrapolated the 3-million-year-old chronology of reversals, determined by Cox, Dalrymple, Doell, MacDougall, Chamalaun, and Tarling, back to 80 million years. They assumed that seafloor in the South Atlantic Ocean, between South America and Africa, had formed at a constant rate, and therefore that Africa and South America had drifted apart at a constant rate, at least for the past 80 million years. Even many confirmed drifters who accepted seafloor spreading and the Vine–Matthew Hypothesis were unwilling to take Heirtzler's time scale seriously. Of course, this extrapolation could be readily tested, if only the ocean floor could be dated. Yet, all but the youngest oceanic basement is buried by a perpetual 'snowfall' of sediment falling down through the ocean to its floor, and, even where a sample from the basement rock could be obtained, dating it reliably had proved to be very difficult.

Deep-sea drilling: a test of the Vine–Matthews Hypothesis

One key assumption and a technological advance circumvented these problems. The dating problem could be resolved by assuming that almost immediately after its formation, new seafloor was covered by biogenic sediment, dead plankton, that sank to the bottom of the ocean. Then, fossils of that plankton found just above the basaltic rock would yield an age that should be nearly the same as that of the rock itself. Working largely with sediment deposited on the ocean floor but now exposed on land, micro-palaeontologists, scientists who study fossils of microorganisms like fossil plankton, had determined when various organisms lived. Thus, given an assemblage of fossil organisms, they could assign an age with an error of 2–3 million years (in the 1960s and for plankton that lived in the past 80 million years, but less than 1 million years today). The problem became: How does one obtain sediment just above

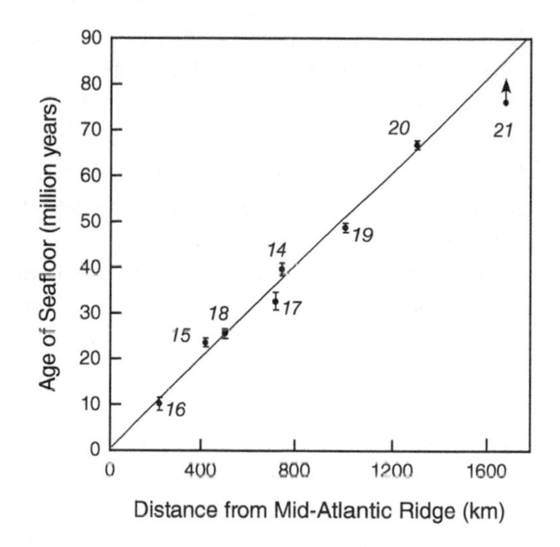

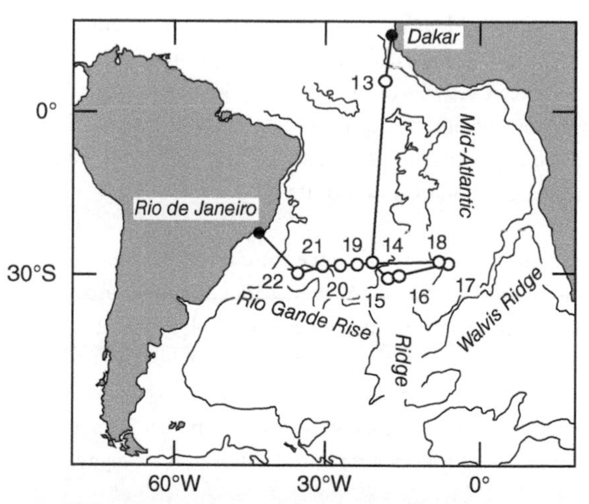

11. South Atlantic drilling results. Bottom: map of the deep-sea drilling ship's track and sites where holes were drilled into sediment. Top: ages of seafloor, with vertical bars showing uncertainties, plotted versus distance from the Mid-Atlantic Ridge in the South Atlantic.

the basalt that was magnetized according to Vine and Matthews, after that basal sediment was buried by hundreds of metres of younger sediment?

New technology came with the deep-sea drilling ship, *Glomar Challenger*, and a large (that is, expensive), multi-institutional, multi-investigator, and eventually multi-national programme to fund its operation, the Deep-Sea Drilling Project. Not only could the *Challenger* drill in the deep ocean through up to 5 km of sediment and into the rocky basement below, but she also could take samples of that sediment along the way. The *Challenger* set sail on her maiden voyage in late 1968. After a few tests and some preliminary holes, she left for the South Atlantic to test the seafloor spreading hypothesis. Even the most confident advocates of seafloor spreading expected some complicated pattern to emerge, and merely hoped that the age of the seafloor would increase outward from the ridge in some way. No one was quite prepared for the beautiful, simple, nearly constant increase in age with distance that was actually found (Figure 11). (My father, J. P. Molnar, an experimental physicist, told me that from the point of view of an experimental scientist, nothing was more beautiful than a straight line of data points.)

The increase in age with distance from the ridge crest in the South Atlantic, as Heirtzler and colleagues had assumed, proved to be remarkably constant. A number of the shipboard scientists had been 'fixists', those who thought that continents had not drifted, when they boarded the *Glomar Challenger* at Dakar, Senegal, for this cruise. When they docked in Rio de Janeiro, Brazil, they left the ship thoroughly converted, exuberant in their joyous recanting.

Fits and starts

From Vine and Matthews's proposing that strips of oppositely magnetized rock would form parallel to mid-ocean ridges to the application of their hypothesis to the dating of ocean floor, progress came rapidly, but not without mistakes. When corrected, however, some of these mistakes brought new understanding and further confirmation of seafloor spreading.

When Cox, Dalrymple, Doell, MacDougall, Chamalaun, and Tarling first found reversals of the Earth's magnetic field, they named them after famous scientists who had contributed to the study of the Earth's field. Then, when they found more intervals with normal or reversed field, and of short duration (*c.*100,000 years), they named them after the locations where these short intervals were recognized. When, from profiles of magnetic anomalies, Pitman, Herron, and Heirtzler inferred countless reversals of the Earth's field, they simply numbered the more obvious anomalies. Anomaly 1 spans the axis of the mid-ocean ridges. Anomaly 2 overlies the flanks of the ridge axes. Anomaly 3, called four-finger Jack by aficionados, consists of four narrow strips of normal polarity, separated by narrow strips of reversed polarity and bounded by wider strips of reversed polarity. Each of the numbered magnetic anomalies has a character of its own. To number the reversals, Pitman, Herron, and Heirtzler and colleagues chose ships' tracks that not only spanned the entire widths of oceans, but also showed clear magnetic anomalies. In their haste, however, they used as a representative profile one that had crossed a large seamount, a submerged mountain. As it turns out, this seamount was strongly magnetized, and it created a large perturbation to the strength of the field near it, a magnetic anomaly. Anomaly 14 was a mistake. (One cannot help but wonder what the superstitious would say if they had numbered another one between anomalies 1 to 13, and the mistake had been anomaly 13.)

In 1965, clear strips of normally and reversely polarized magnetization had been mapped along some small portions of the mid-ocean ridge system, but not quite all of the reversals of the field in the past 3 million years had been identified. In an effort to demonstrate the success of the Vine–Matthews

Hypothesis, Vine himself and Tuzo Wilson of the University of Toronto, who figures prominently in Chapter 3, correlated the magnetic anomalies with the history of the Earth's field, as it was known at that time. They associated the positive anomaly over ridge axis with the current period of normal polarity of the Earth's magnetic field, thought in 1965 to have prevailed for the last 1 million years. They associated the adjacent negative anomaly with the older period of reversed polarity, the next positive anomaly farther from the ridge axis with the previous period of normal polarity, and so on. Then on one axis they plotted the distance from the ridge crest to the changes from positive to negative anomaly, or negative to positive anomaly, and on the other axis the corresponding ages of reversals, from normal to reversed, or reversed to normal, as they were known at the time (Figure 12). Of course, the anomalies associated with the older reversals lay farther from the ridge axis than the younger ones. The ratio of the distance between changes from negative to positive and then positive to negative anomalies divided by the duration of the corresponding interval of normal polarity should equal the rate at which the underlying seafloor formed. Vine and Wilson inferred that the average rate that seafloor formed at the Juan de Fuca ridge, just off the west coast of the USA, varied from as slow as 6 mm per year to as fast as 25 mm per year (Figure 12).

Imagine Vine and Wilson's surprise, if not glee, when shortly after their paper was published, Doell and Dalrymple published a paper demonstrating that the Earth's field underwent a short period of reversed polarity near ~800,000 years ago, preceded by another short period of normal polarity (the Jaramillo Event). This meant that Vine and Wilson had assigned incorrect ages to all of the correlations of switches between positive and negative anomalies with normal and reversals of the Earth's field. Each switch had been assigned an age too old, but by differing amounts (Figure 12). The corrections to the ages differed because of the random occurrence of reversals and the different durations of each

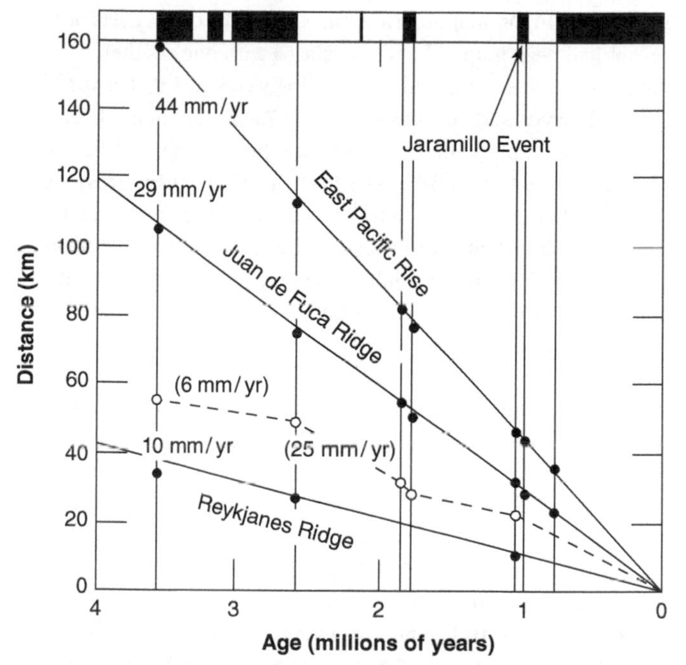

12. Ages of reversals in the magnetic field, spreading rates, and Vine and Wilson's error. Horizontal axis gives ages of reversals (black indicates normal polarity and white reversed) and vertical axis gives distances from axes of three different mid-ocean ridges. Open circles show the ages that Vine and Wilson assumed for the Juan de Fuca Ridge, just west of the northwestern USA, and before the Jaramillo event was identified. With the Jaramillo event included, all of Vine and Wilson's ages are shifted to become younger, so that with correct ages, distances versus ages define straight lines, which means that the seafloor formed at each ridge (or rise) was created at a constant rate.

period of normal or reversed polarity. The new correlations, however, made much more sense. They vanquished the irregularities in the rate at which seafloor had seemed to be created; the spreading rate for the past 3 million years no longer appeared to be erratic, but now was virtually constant, at 29 mm per year. Nature can be kind.

By the early 1970s, magnetic anomalies over the oldest parts of the oceans had been mapped and correlated with one another, but the correlations were not overwhelmingly convincing. For the past 200 million years, seafloor beneath the Pacific Ocean has been created more rapidly than that beneath the Atlantic, and seafloor as old as 120 to 150 million years had moved long distances since it formed. In the northern hemisphere, where the Earth's field currently points downward (and northward), seafloor that formed during normal polarities would also be magnetized downward, but in the southern hemisphere seafloor of the same age would be magnetized upward (but also northward). Suppose seafloor formed in the southern hemisphere, but then moved to the northern hemisphere. Magnetometers again would record positive anomalies over seafloor with magnetization pointing downward, but now such positive anomalies (over seafloor formed in the southern hemisphere but now lying in the northern hemisphere) should be associated with periods of reversed, not normal, polarities of the Earth's magnetic field. In 1972, Roger Larson, then also at Lamont Geological Observatory, and Walter Pitman recognized that seafloor in the western North Pacific had, in fact, formed thousands of kilometres south of where it currently lies, and south of the equator. With the recognition of that fact, the previously unconvincing correlations of magnetic anomalies older than 120 million years could be rejected, and a new, convincing set of correlations emerged clearly. Moreover, with an approximate knowledge of the history of the Earth's field in that period, if much less precisely known than for the past 3 million years, Larson and Pitman could date virtually all of the seafloor.

Deepening of seafloor

If seafloor is made at mid-ocean ridges, and one side moves away from the other, why is the seafloor shallow at the ridges? A better question might be: Why does the seafloor become deeper at increasing distances from the ridges? Plate tectonics provides a simple answer.

Shortly after its birth, the life of an ordinary piece of seafloor that has been created at a mid-ocean ridge becomes rather dull, like nearly every other piece of seafloor. Although the amounts and types of sediment deposited on the ocean bottom vary from place to place, the composition and structure of the oceanic crust is remarkably uniform beneath the deep ocean. The structure of oceanic lithosphere depends primarily on its age, the time that has elapsed since it formed at a mid-ocean ridge. New lithosphere steadily cools by losing heat to the overlying ocean as it moves away from the spreading centres. Near the mid-ocean ridges, the lithosphere is relatively hot and thin (Figure 5), and heat flows rapidly through the crust and into the overlying ocean. As the lithosphere ages, it thickens, and the rate at which it cools decreases.

Recognizing this in 1967, Dan McKenzie of Cambridge University developed a simple physical and mathematical description of how the lithosphere cools. His target was abundant, if noisy data that showed that the rate that heat is lost through the seafloor decreases with the age of lithosphere. Like many profound ideas, McKenzie's equation describing thermal structure proved to be much more valuable than he had anticipated, because with modest extension, it could also account for the depth of the ocean floor. As the lithospheric plate loses heat and cools, like most solids, it contracts. This contraction manifests itself as a deepening of the ocean.

In the early 1970s, John Sclater and Jean Francheteau at Scripps Institute of Oceanography showed that when the depths of the various oceans were plotted as a function of the age of the seafloor, one curve, now called the 'Sclater curve', could describe the deepening of the oceans from about 2.5 km at the ridge axis to about 6 km in the oldest parts of the oceans. Sclater, however, was frustrated by an inability to extend McKenzie's mathematics to account for the depths. One Friday afternoon in 1971, a young undergraduate student at the University of California at San Diego,

Miller Lee Bell, probably tired of doing tedious work for Sclater, took the equations home for the weekend and returned with the solution on Monday. By giving the observations a simple, quantitative physical interpretation, Sclater, Roger Anderson, and Bell could then use the observed depths of the seafloor of different ages to infer parameters such as the thickness of the lithosphere that entered into McKenzie's theory. As occurred many times with plate tectonics, it offered a quantitatively more accurate description than had initially been expected.

With magnetic anomalies to date the seafloor and Sclater's curve relating depth and age, the difference between the Atlantic, with its 'ridge', and the Pacific and its 'rise' became comprehensible. Seafloor spreading in the Pacific occurs two to five times faster than it does in the Atlantic. Thus, the relatively narrow Mid-Atlantic Ridge is underlain by seafloor the same age as the wider East Pacific Rise. Moreover, we now understand that when seafloor spreading is slow, new basalt rising to the surface at the ridge axis can freeze onto the older seafloor on its edges before rising as high as it would otherwise. As a result, a valley, Ewing and Heezen's rift valley, forms. Where spreading is faster, however, as in the Pacific, new basalt rises to a shallower depth and no such valley forms. The differences between the floors of the Pacific and Atlantic ceased to be obstacles to understanding and became examples of general processes.

Sea level

With a theory for predicting the depths of oceans, the history of sea-level changes became amenable to simple analysis. Geochemists are confident that the volume of water in the oceans has not changed by a measurable amount for hundreds of millions, if not billions, of years. Yet, the geologic record shows several periods when continents were flooded to a much greater extent than today. For example, 90 million years ago, the Midwestern United States and neighbouring Canada were flooded. One could have sailed due north from the Gulf of Mexico to Hudson's Bay and into the Arctic. If we use New Orleans or St Louis as a reference, we find that sea level was roughly 250 m higher than it is today. If sea level has risen and fallen, while the volume of water has remained unchanged, then the volume of the basin holding the water must have changed.

The rates at which seafloor is created at the different spreading centres today are not the same, and such rates at all spreading centres have varied over geologic time. Imagine a time in the past when seafloor at some of the spreading centres was created at a faster rate than it is today. If this relatively high rate had continued for a few tens of millions of years, there would have been more young ocean floor than today, and correspondingly less old floor (Figure 13). Thus, the average depth of the ocean would be shallower than it is today, and the volume of the ocean basin would be smaller than today. Water should have spilled onto the continent. Most now attribute the high sea level in the Cretaceous Period (145 to 65 million years ago) to unusually rapid creation of seafloor, and hence to a state when seafloor was younger on average than today.

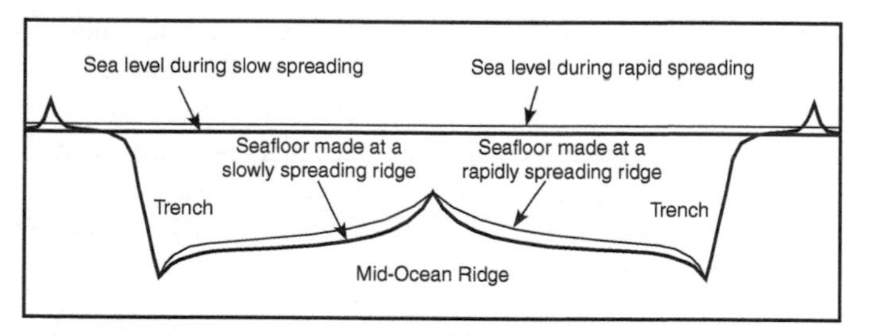

13. Simple cartoon illustrating how changing rates of seafloor spreading can affect sea level. Dark and fine lines show bathymetric profiles for slow and more rapid spreading. For faster spreading the shallower depths of water would make water spread onto the surrounding continents (causing sea-level rise).

Chapter 3
Fracture zones and transform faults

When Hess exploited the work of Ewing, Heezen, and others to propose seafloor spreading, and concurrently Menard, Dietz, and colleagues mapped fracture zones in the Pacific (Figure 14), the relationship of one to the other was not obvious. Fracture zones were a feature of the Pacific, but not yet recognized in the Atlantic. Then, by the mid-1960s and largely through the work of Bruce Heezen and Marie Tharp at Lamont Geological Observatory, fracture zones in the Atlantic Ocean had been mapped, and they clearly showed something that Menard and Dietz could not have seen: offsets in the crest of a ridge, the Mid-Atlantic Ridge, like those in Figure 15. The spreading apart of two plates along a mid-ocean ridge system occurs by divergence of the two plates along straight segments of mid-ocean ridge that are truncated at fracture zones. Thus, the plate boundary at a mid-ocean ridge has a zig-zag shape, with spreading centres making zigs and transform faults making zags along it.

In 1965, J. Tuzo Wilson of the University of Toronto and someone with an established reputation for bold thinking, saw how to make sense of Menard and Dietz's fracture zones in the Pacific and Heezen and Tharp's in the Atlantic. He pointed to the simple difference between the implicit view that most had taken, and another that incorporated seafloor spreading. Most saw slip as occurring along the entire fracture zone, and therefore along a

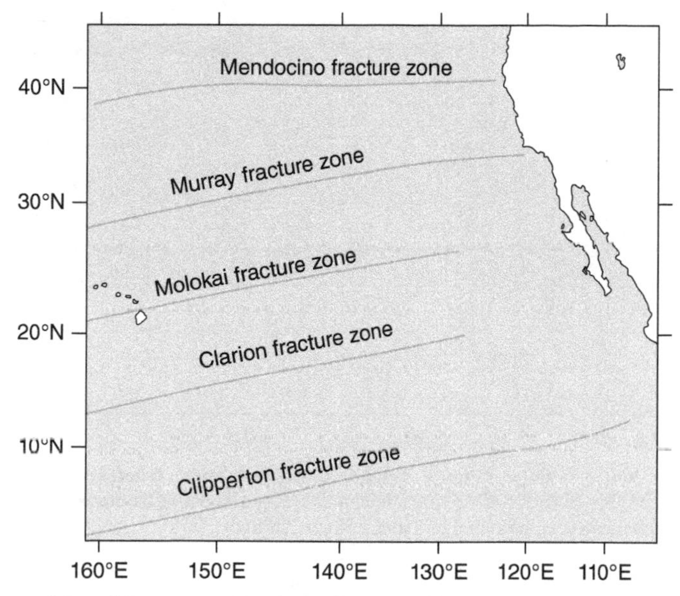

14. Map of the eastern North Pacific, west of the western USA and Mexico, showing the traces of fracture zones.

strike-slip, or 'transcurrent', fault (Figure 16, bottom). Wilson proposed instead transform faulting, for which slip actively occurs only along a segment of the fracture zone at any time; that segment is the portion of the fracture zone that lies between adjacent segments of mid-ocean ridge. As slip occurs, the inactive portions of the fracture zone (dashed lines in Figure 16, top) grow longer.

We now know Wilson's insight to have been squarely on the mark. Students taught transform faulting today sometimes wonder with disbelief how such an obvious inference was not recognized sooner. So, before discussing how Wilson's transform faulting was tested, let's look back on the common view of strike-slip, or transcurrent, faulting before 1965.

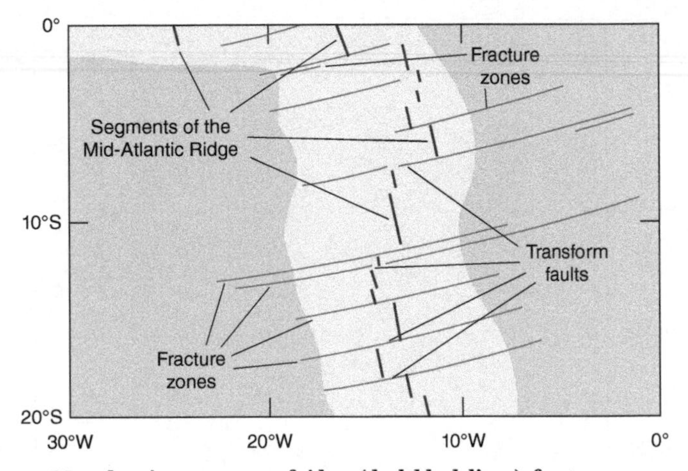

15. Map showing segments of ridges (dark black lines), fracture zones (long thin black lines), and transform faults (portions of fracture zones between segments of ridges) in the South Atlantic.

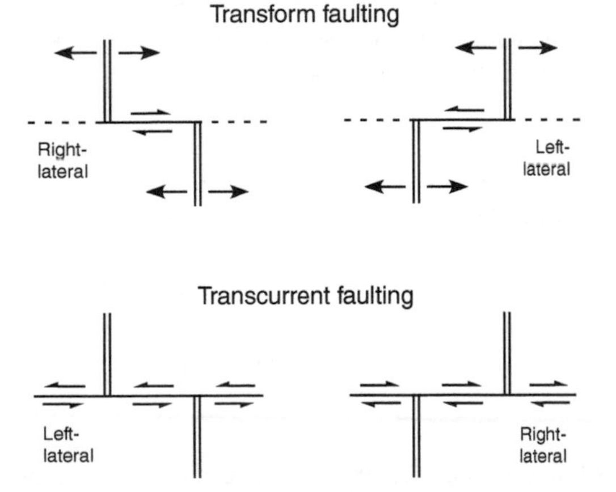

16. Transform (top) and transcurrent faulting (bottom) at fracture zones that offset mid-ocean ridges. Top: only the segment of the fracture zone between segments of ridge crest is active. Bottom: with transcurrent faulting, the entire fracture zone would be active, and the present-day offset of the segments of ridge crests would be growing with time, as slip occurred along the entire fracture zone.

Fracture zones

As noted in Chapter 1, Menard and Dietz had mapped long stretches, hundreds of kilometres, of relatively flat smooth seafloor that would suddenly give way to rough topography. Moreover, on opposite sides of the rough topography the mean depth of the ocean commonly differed by hundreds of metres. Menard and Dietz had focused on the linearity of the Mendocino fracture zone (Figure 14), but with more cruises, Menard realized that loci of rough seafloor were aligned along nearly straight lines on a map, nearly great circles on a globe. Although Menard, Dietz, and colleagues had only a few observations, because ships' crossings of many of the fractures zones were widely spaced, Menard took a bold step in assuming that the fracture zones, which are now obviously continuous features, were in fact so.

For example, in 1956, the *Spencer F. Baird*, an oceanographic research vessel, was sailing westward from south of Mexico near the Clipperton fracture zone (Figure 14). The Chief Scientist, Robert (Bob) Fisher of Scripps Institute of Oceanography, awoke in the morning and asked the captain where the ship was. Extrapolating from bathymetry farther east, Fisher deduced that for the ship to be where the captain said, it should have crossed the Clipperton fracture zone. Eagerly examining bathymetric profiles from the previous night, however, he found records of only relatively smooth ocean floor. He then informed the captain of an error in navigation, a strong and potentially insulting statement to make to the captain of a ship, but on this occasion, fracture-zone linearity trumped a captain's navigational expertise.

Menard inferred that fracture zones might mark great strike-slip faults (Figure 4). The linearity of fracture zones screams 'strike-slip faulting' to geologists. Continued slip parallel to the trend of a fault, as occurs on strike-slip faults, polishes the fault surface to make it straight. All major strike-slip faults, like the San Andreas fault in California, are remarkably straight; pilots flying from Los Angeles to San Francisco are known to navigate simply by following the trace of that fault, which can be seen clearly from above.

Determining the sense of slip on a strike-slip fault, however, requires other observations: does an observer standing on one side of the fault see the opposite side move to the right, 'right-lateral', or to the left, 'left-lateral' (Figure 16)? When geologists measure the amount of slip on such faults, they do so by finding features that once were continuous across the fault and that were subsequently displaced by slip on the fault, like the fence shown in Figure 17 and conspicuously offset by the 1906 San Francisco earthquake, which ruptured the north end of the San Andreas fault.

Not long after ships first towed magnetometers, and a few years after Menard and Dietz had discovered fracture zones, Victor Vacquier, Arthur Raff, and Robert Warren, then at Scripps Institute of Oceanography too, showed that recognizable magnetic anomalies were also offset along fracture zones (Figure 18). They sailed their ship parallel to the Mendocino fracture along courses north and south of the fracture zone, and they observed the same sequence of magnetic anomalies on both sides, except that one sequence was found 1140 km east of the other (Figure 18). Moreover, they found offsets of sequences of magnetic anomalies across all of the fracture zones in the Pacific.

Building from the common sense image associated with earthquakes (Figure 17), Vacquier, Raff, and Warren's discovery that the magnetic anomaly pattern north of the Mendocino fracture zone was repeated 1140 km to the east just south of the fracture zone (Figure 18) provided evidence that was what any geologist might expect for a major strike-slip fault. That 1140-km offset was immediately, and logically, cited as evidence for 1140 km

17. G. K. Gilbert's photograph of a fence offset in the San Francisco earthquake of 1906. The fence receding into the distance was displaced to the right relative to its continuation in the foreground on the left.

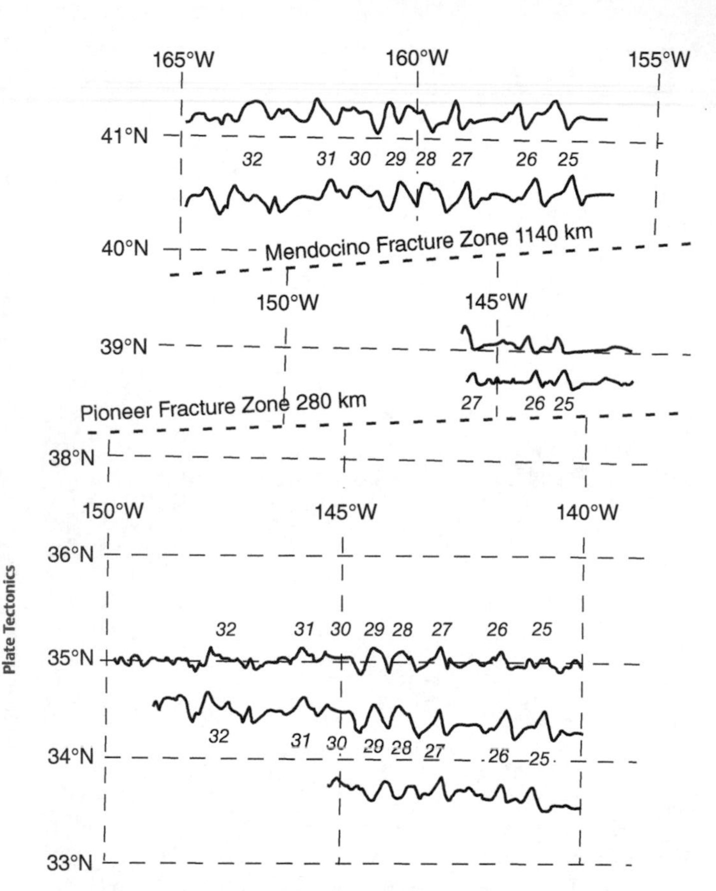

18. Profiles of magnetic anomalies (wiggly lines) plotted along ships' tracks in the North Pacific Ocean. Short dashed lines indicate where fracture zones had been mapped. Magnetic anomalies 25 to 32 south of the Pioneer Fracture Zone lie 1420 (= 1140 + 280) kilometres east of those north of the Mendocino Fracture Zone.

of left-lateral strike-slip motion on the Mendocino fracture zone; they inferred that the seafloor north of the fracture zone had been displaced 1140 km westward relative to that south of the fracture zone and 1420 km relative to those farther south.

Menard, Dietz, and Fisher had not only mapped long linear scars in the seafloor, but they also noticed that bathymetric contours were offset; those contours of constant depth on the smooth seafloor on one side of a fracture zone, and aligned approximately perpendicular to it, were systematically offset by hundreds to a thousand kilometres to a position on the opposite side of the fracture zone where they continued again perpendicular to the fracture zone. At first, this step in topography led them astray, but later it was recognized as another symptom of strike-slip faulting, consistent with the offset magnetic anomalies. Imagine a gently sloping roof on a building with a strike-slip fault cutting through the building and its roof. Strike slip would not raise the roof on one side of the fault relative to that on the other. Nevertheless, if we walked along the roof maintaining a constant height, when we reached the fault, the roof on the opposite side of the fault would be either higher or lower. The portion at the same height would be displaced horizontally from where we stand at the fault, and a part of the roof that had been higher or lower, before slip occurred, would have been displaced to lie in front of us. Menard and Dietz quite sensibly interpreted the displaced bathymetric contours as additional evidence for strike-slip faulting.

These geologists simply exploited an understanding of what seemed obvious—strike-slip faulting—and then they inferred that the bathymetric contours and the magnetic anomalies had once been aligned, and later were offset by slip along the fracture zones—the same logic that we apply to slip during earthquakes and offsets of fences that cross the fault (Figure 17). The fracture zones could not be traced on shore, and therefore some kind of *deus ex machina* was required for them to end abruptly at the continental margins. Nevertheless, the logic of such an analogy with offsets associated with earthquakes (Figure 17) was too simple and reasonable to be discounted easily, or for an alternative even to be considered.

We now understand the absence of eastward continuations of the fracture zones on land: because of the tectonic evolution of this region, most segments of the spreading centres at which the North Pacific Ocean floor had been created no longer exist. The seafloor that was created on the east side of these spreading centres has slid beneath western North America along a subduction zone that followed the west coast of North America. Most of that subduction zone no longer exists; only a short segment along the coasts of northern California, Oregon, Washington, and southern British Columbia remains active. Because of this history of subduction beneath western North America, no active transform fault can be found along the huge fracture zones in the eastern Pacific, and they do not offset a pair of active spreading-centre segments as they do in the Atlantic (Figure 15). As a result, the intimate relationship between fracture zones and the mid-ocean ridge system, found later in the Atlantic, was not apparent in the Pacific.

In retrospect, it appears that Nature played a dirty trick, for now we understand that the offset of the anomalies that Vacquier, Raff, and Warren had discovered reflects the former existence of a transform fault that lay between spreading-centre segments that were 1140 km apart, and at which magnetic anomalies and seafloor were forming.

The recognition of transform faulting

The recognition that transform, not transcurrent, faulting had occurred required the mind of a scientist who was always in search of simple ideas, unencumbered by complexity. To persuade others it helped that Tuzo Wilson was somewhat of a showman. In the summer of 1965, a number of marine geophysicists gathered in Ottawa to report and review their recent work. In the middle of a serious session, Wilson rose to give his scheduled lecture. With a twinkle in his eye, according to Tanya Atwater, a graduate student at that time and about whom more is written below, he passed out sheets of paper to everyone in the room. The papers had a few dashed and dotted lines that said, 'cut here', or 'fold here'. Everyone laughed, of course, thinking, 'What a way to do science!' Wilson's goal was to give everyone a toy that he or she could play with and understand what he was saying.

He reasoned that if seafloor spreading were to occur and the Vine–Matthews Hypothesis were correct, then the sense of motion along the fracture zones would be opposite to that of transcurrent faulting (Figure 16, bottom), the sense of slip that associated segments of mid-ocean ridges with offset fences, roads, rows of trees, etc. during earthquakes (Figure 17). Active faulting on the fracture zones would be confined to the segments between the ridge crests, and the segments on the flanks of the ridge would be inactive (Figure 16, top). Thus the horizontal distance between ridge crests, the bathymetric contours, and the magnetic anomalies would be essentially permanent features in the ocean floor and would not equal the amount of displacement that accumulated over time by slip along the fracture zone.

In presenting the idea, Wilson focused on the two major differences between ordinary strike-slip faults, or transcurrent faults, and transform faults on fracture zones. (1) If transcurrent faulting occurred, slip should occur along the entire fracture zone; but for transform faulting, only the portion between the segments of spreading centres would be active. (2) The sense of slip on the faults would be opposite for these two cases: if right-lateral for one, then left-lateral for the other (Figure 16). Rarely does science present a simple, yet fundamental, idea with such an easy method for testing it.

The occurrences of earthquakes along a fault provide the most convincing evidence that the fault is active. Slip on most faults and most deformation of the Earth's crust to make mountains occurs not slowly and steadily on human timescales, but abruptly during earthquakes. Accordingly, a map of earthquakes is, to a first approximation, a map of active faults on which regions, such as lithospheric plates, slide past one another (Figure 3).

Slip on faults during earthquakes

In 1965, Lynn Sykes, a recent PhD graduate at Lamont Geological Observatory, read Wilson's paper. Sykes had been carrying out a systematic study of the earthquakes in various regions of the Earth. When he read Wilson's paper, he knew that nearly all earthquakes along the mid-ocean ridge system occurred either on the spreading centres or in the segments between them along the fracture zones (Figure 19). Nevertheless, it was not until the following year, when Walter Pitman and Jim Heirtzler showed him the symmetry of magnetic anomalies and confirmation of the Vine–Matthews Hypothesis, that Sykes dropped his other research and set off to test Wilson's idea. A convincing demonstration of transform faulting required the determination of the sense of displacement on the fault, the difference between the cases shown in Figure 16 (or something yet different from both).

When an earthquake occurs, slip on a fault takes place. One side of the fault slides past the other so that slip is parallel to the plane of the fault; the opening of cracks, into which cows or people can fall, is rare and atypical. Repeated studies of earthquakes and the surface ruptures accompanying them show that the slip during an earthquake is representative of the sense of cumulative displacement that has occurred on faults over geologic timescales. Thus earthquakes give us snapshots of processes that occur over thousands to millions of years.

Two aspects of a fault define it: the orientation of the fault plane, which can be vertical or gently dipping, and the sense of slip: the direction that one side of the fault moves with respect to the other (Figure 4). Obviously, if one stands on one side and detects movement of the other to the right, upward and northward, then

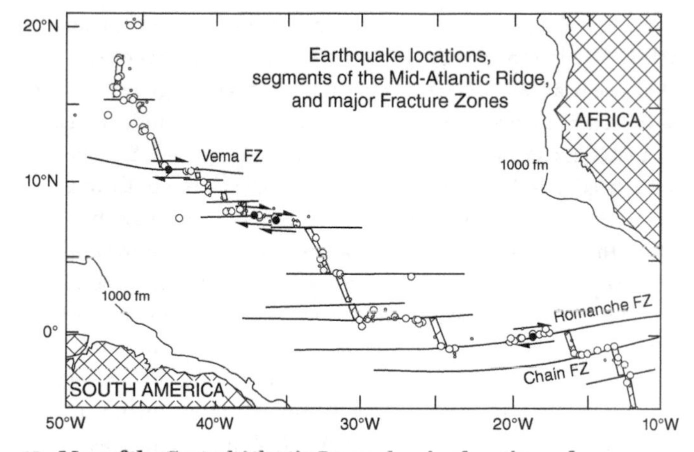

19. Map of the Central Atlantic Ocean showing locations of earthquakes (open and dark circles), segments of ridge axes (double lines), and fracture zones (dark black lines). Arrows surrounding dark circles show the sense of slip that Lynn Sykes determined for these earthquakes.

if one stands on the other side of the fault, the movement will also be to the right, but downward and southward.

To a first approximation, boundaries between plates are single faults. Thus, if we can determine both the orientation of the fault plane and the sense of slip on it during an earthquake, we can infer the direction that one plate moves with respect to the other. Often during earthquakes, but not always, slip on the fault offsets the Earth's surface, and we can directly observe the sense of motion, as in Figure 17. In the deep ocean, however, this cannot be done as a general practice, and we must rely on more indirect methods. Such methods exploit seismic waves, which propagate from earthquakes and explosions through the Earth to seismographs now deployed over the Earth.

When people feel an earthquake, usually they feel the waves radiated by it, not the slip of one side of the fault past the other.

Two types of waves pass through the interior of the Earth, P and S waves. The first to arrive are P (once called 'primary') waves, which we hear as sound waves when they pass through liquids and gases. When a P wave propagates through a solid, liquid, or gas, the material contracts and expands as the wave passes. S (once called 'secondary') waves travel only through solids, not through liquids or gases; they require elastic resistance to changes in shape, the kind of resistance that allows a diving board to support a diver on the end of it. When an S wave passes through a solid, the solid changes shape, but there is no contraction and expansion of material. Seismologists use P and S waves to determine the structure of the Earth, as well as to study earthquakes themselves. Initially, the determination of the nature of faulting during an earthquake using seismic waves, called 'fault plane solutions', relied exclusively on recordings of P waves, but now with networks of accurately calibrated seismographs, entire seismograms, not just the first arriving P wave, can be used to determine the nature of faulting during an earthquake. Here, however, we will confine ourselves to a discussion of P waves.

When an explosion occurs, the material surrounding the explosion initially moves away from the explosion (Figure 20). When an observer first senses the P wave from an explosion, the medium surrounding the observer, and also the observer, him- or herself, will move away from the explosion. We say that the 'first motion of the P wave' is away from the source and is 'compressional', because the medium on which the observer stands first compresses with the passage of the wave before expanding back to its original volume. For an implosion the material surrounding the source moves toward it, and so is the first motion of the P wave from an implosion. The material near the observer initially expands, or dilates, and the first motion of the P wave is said to be 'dilatational'.

During an earthquake, slip occurs on a fault without a change in volume. In some directions from the earthquake, the material in the immediate vicinity of the fault, and on the same side of the

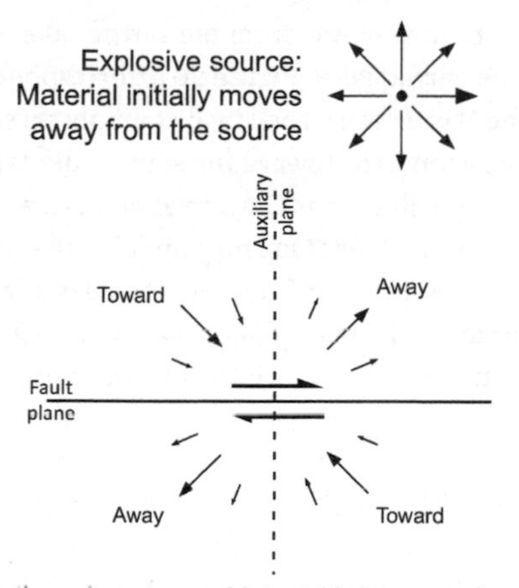

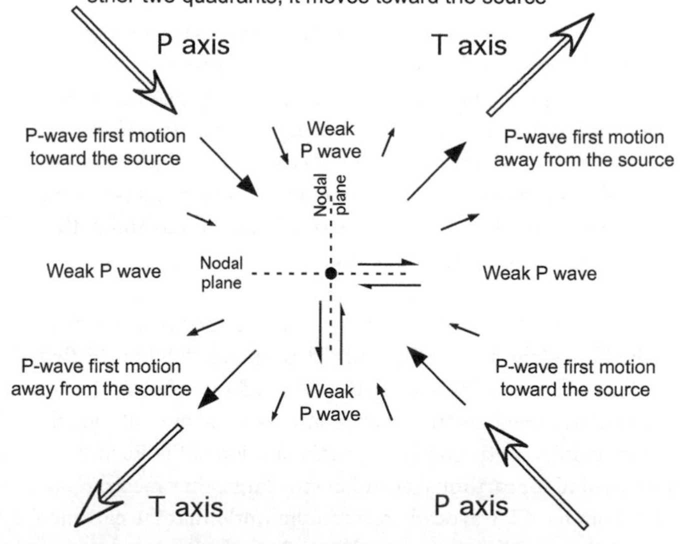

20. Predicted P-wave first motions from an explosion (top) and an earthquake (middle) and observed first motions (bottom). Dark arrows show the direction of the first motion of a P wave radiated by explosions and earthquakes, with their lengths indicating strengths of P-wave signals.

fault as the observer, moves away from the earthquake; elsewhere, it moves toward the earthquake (Figure 20). Correspondingly, the first motions of the P waves recorded by distant observers are away from (compressional) or toward the source (dilatational). In theory, an earthquake will send no P wave at all in directions that lie within the plane of the fault. One may imagine that the P waves with opposite polarities and sent from the two sides of the fault will cancel one another. The fault plane is said to be a 'nodal plane' in the radiation pattern of P waves. In theory, the earthquake also sends no P wave in any direction perpendicular to the direction of slip on the fault, because material moves neither toward nor away from the earthquake. This family of directions along which P waves ought not to emanate from the source defines another nodal plane, the 'auxiliary plane'. It follows that earthquakes send weak P-wave signals in directions near the nodal planes.

Because these two nodal planes are perpendicular to one another, in three dimensions the radiation pattern consists of four quadrants separated by two nodal planes. These quadrants are analogous to huge segments of an orange or grapefruit, but so large that there are only four segments and their edges intersect each other at angles of 90°. In two opposite quadrants (segments), the first motions are toward the source (dilatational) and in the other two, away from it (compressional).

To determine the orientation of the fault and the sense of slip on it, the fault plane solution, one first determines whether the first motions of P waves from an earthquake and recorded by stations in different directions from it are compressional or dilatational. Seismographs continuously record the motion of the Earth's surface, and for earthquakes sufficiently large, they record clear first motions of P waves wherever in the world that the earthquake occurred, except for seismographs to which the P waves leave the source near a nodal plane. With a sufficient number of observations to define four quadrants (the four segments of our orange or grapefruit), we determine the orientations of two planes, one of which is the plane that ruptured in the earthquake. The other, the auxiliary plane, is oriented perpendicular to the direction that one side of the fault moved with respect to the other side.

To use fault-plane solutions, there is one more stage of interpretation. We can reliably predict the first motions of P waves (and S waves too) and the orientations of the two nodal planes, if we know the orientation of the fault and the direction that one side moves with respect to the other on it. Given only the orientations of the nodal planes, however, we cannot decide which nodal plane is the plane on which slip occurred during the earthquake. In Figure 20, if the auxiliary plane were, in fact, the fault plane, and the fault plane were the auxiliary plane, the observed P-wave first motions would be the same as those shown. Although in theory a careful analysis of the seismic waves can resolve the ambiguity of which plane is the fault plane, in practice such analysis is difficult and tedious, and results are only rarely convincing. Analyses of the first motions of S waves, or of other parts of the seismograms, in general, do not help resolve this ambiguity. Instead one infers the fault plane from the geologic structure of the source region and then deduces the sense of slip on it. If one of the nodal planes, but not the other, lies parallel to a known fault or to a planar zone of earthquakes, one logically assumes this plane to be the fault plane.

In the case of earthquakes on fracture zones and transform faults, invariably both nodal planes are nearly vertical, and the intersection of one with the Earth's surface makes a line, the 'strike' of the fault, that is parallel to the fracture zone. When Sykes first observed this, he assumed that the earthquake occurred by slip on a fault parallel to the fracture zone. Then from the first motions of the P waves in the different quadrants, he demonstrated that transform, not transcurrent, faulting had occurred (Figure 19). Probably no single study persuaded more seismologists that seafloor spreading occurs than Sykes's study of fault plane solutions of earthquakes on fracture zones.

As discussed more in Chapter 4, fault plane solutions of earthquakes have helped us understand other plate boundaries, not just transform faults.

Different types of transform faults

The discussion above concentrated on transform faults that connect segments of spreading centres. Other transform faults can exist as well, such as between two subduction zones, as shown on the left side of the block diagram in Figure 5, or even between a spreading centre and a subduction zone. Nearly all of these exist in the real world, but usually they are not as simple as ridge–ridge transform faults. In terms of plate tectonics, transform faults are plate boundaries along which lithosphere is neither created nor destroyed; they terminate at other types of plate boundaries or at 'triple junctions', where three plates meet.

Chapter 4
Subduction of oceanic lithosphere

Because seafloor spreading creates new seafloor at the mid-ocean ridges, the newly formed crust must find accommodation: either the Earth must expand or lithosphere must be destroyed at the same rate that it is created. When Xavier Le Pichon, then at Lamont Geological Observatory, calculated how the major plates have moved relative to one another (see Chapter 5), he found that if lithosphere were not subducted into the athenosphere, the Earth would be expanding into an asymmetric, disc-like object (or would have begun as a cigar-shaped object for it to have grown into its present nearly spherical shape). The network of mid-ocean ridges, most of which trend roughly north–south, would make the equatorial region grow rapidly with little divergence of material at the poles. Abundant evidence, however, shows that subduction of lithosphere does occur.

Years before plate tectonics was recognized, advocates of continental drift had imagined convective flow of rock within the mantle (like ultra-slow movement of boiling water), and they had concluded that downward flow took place beneath 'island arcs', arcuate belts of volcanic islands like those forming the Lesser Antilles in the Atlantic, the Aleutian, Kurile, and Mariana Islands of the Pacific, or the islands of Sumatra, Java, and Flores of Indonesia in the Indian Ocean. It was also apparent to those far-sighted scientists that similar structures characterize the

Andes of South America or the Cascade volcanoes of the northwestern United States, despite the absence of the arcuate chain of islands. In the late 1960s, a plausible image of the structure of such regions was developed, and with it a more precise understanding of the processes taking place was put forth.

The principal ingredient in this new image in the 1960s, absent in most of the earlier conceptions, was the lithosphere. Whereas earlier views portrayed diffuse sinking of material, the new image centred on the strong lithosphere (Figure 5). A cold, dense, strong slab of oceanic lithosphere, approximately 100 km thick, plunges into the sublithospheric mantle at 'island arcs'. This subduction of lithosphere directly or indirectly accounts for essentially all of the basic features associated with 'island arcs' and similar regions like the Andes.

In the 1940s, two eminent seismologists at the California Institute of Technology (Caltech), Beno Gutenberg and Charles Richter (known for his magnitude scale that quantifies 'sizes' of earthquakes), had noted four characteristics common to 'island arcs' (Figure 21): (1) a belt of volcanoes, commonly forming an arcuate chain of volcanic islands and hence the name, but also occurring along the margins of continents (like the Andes or the Cascade Mountains); (2) a deep trench that lies seaward of the island arc and approximately equidistant from the line of volcanoes; (3) a belt of shallow-focus earthquakes (depths less than 70 km) between the trench and the volcanoes; and (4) an inclined zone of intermediate-depth earthquakes (depths between 70 and 300 km), and in some regions also of deep-focus earthquakes (depths between 300 and 700 km), that dips beneath the volcanoes.

To this list we may now add three more characteristics. (5) Slightly shoaled seafloor lies seaward of the trench to form a subtle 'outer

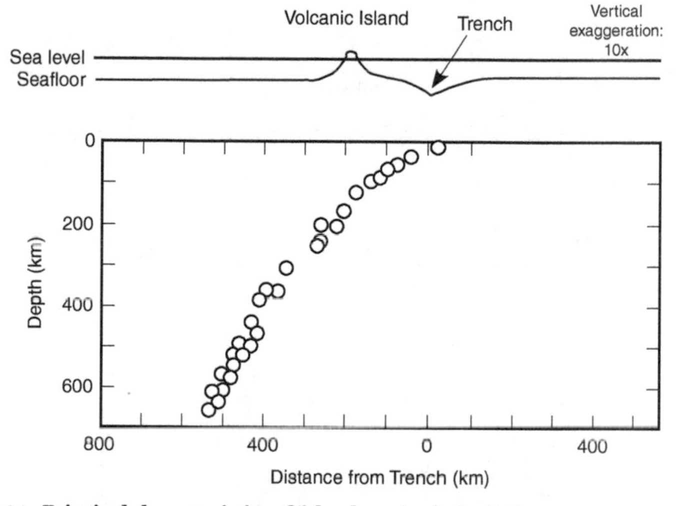

21. Principal characteristics of 'island arcs': a belt of volcanoes, a deep-sea trench approximately 130 km from the volcanoes, a zone of shallow earthquakes (depths < 70 km) dipping beneath the volcanoes, and an inclined seismic zone continuing to greater depth.

topographic rise' (Figure 22). (6) Shallow-focus earthquakes occur in a narrow belt beneath the trench axis, and separate from the zone that dips beneath volcanoes. (7) A thin zone, roughly 100 km in thickness, through which seismic waves propagate at abnormally high speeds and with unusually low loss of energy, follows the inclined zones of intermediate- and deep-focus earthquakes (Figure 23). As discussed below, we now know too that these intermediate- and deep-focus earthquakes occur *within* the zone of high seismic wave speeds and low energy loss.

Although an arcuate chain of volcanic islands does not characterize the continental margins of western South and Central America, Mexico, Alaska, or even Japan, the presence of volcanoes and all of the other features allows them to be defined as 'island-arc structures' too.

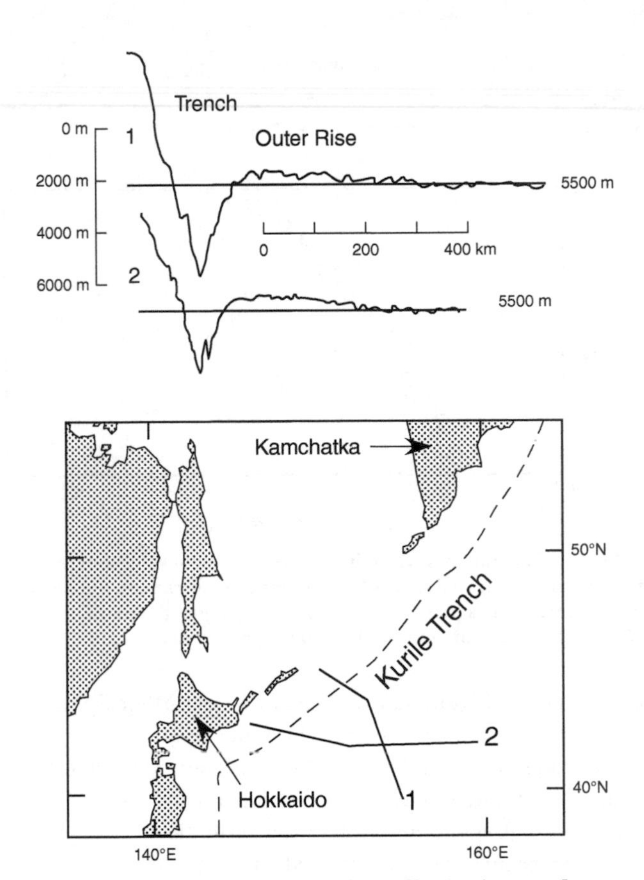

22. Vertically exaggerated bathymetric profiles (top) across deep-sea trenches and outer topographic rises, and map (bottom) showing locations of profiles.

As noted above, for the Earth not to expand (or contract), the sum total of new lithosphere made at spreading centres must be matched by the removal, by subduction, of an equal amount of lithosphere at island arc structures. The subduction process, however, differs fundamentally from that of seafloor spreading, in that subduction is asymmetric. Whereas two plates are created

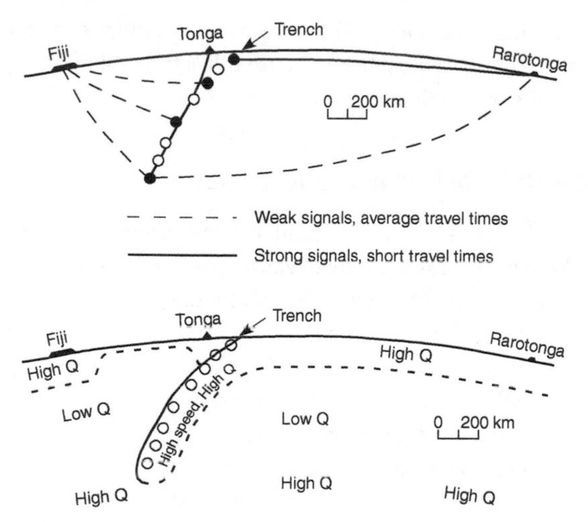

23. Cartoon cross section across the Tonga–Fiji region showing (top) paths for P and S waves with high and low speeds and high and low Q (and therefore low and high attenuation), and the interpretation of these observations (bottom).

and grow larger at equal rates at spreading centers (mid-ocean ridges and rises), the areal extent of only one plate decreases at a subduction zone. The reason for this asymmetry derives from the marked dependence of the strength of rock on temperature.

At spreading centres, hot weak rock deforms easily as it rises at mid-ocean ridges, cools, and then becomes attached to one of the two diverging plates. At subduction zones, however, cold and therefore strong lithosphere resists bending and contortion. When we move a bed into an apartment, moving the flexible mattress around tight corners is much easier than moving the stiff box spring. Similarly, two plates of lithosphere, each some 100 km thick, cannot simply approach one another, turn sharp corners (as mattresses can), and dive steeply into the asthenosphere. Much less energy is dissipated if one plate undergoes modest flexure and then slides at a gentle angle beneath the other, than if both plates were to undergo pronounced bending and then plunged together steeply into the asthenosphere. Nature takes the easier, energetically more efficient, process.

Deep-sea trenches and outer rises

Before it plunges beneath the island arc, the subducting plate of lithosphere bends down gently to cause a deep-sea trench (Figures 21 and 22). (Note that the subducted plate is not strictly rigid, but does deform by a modest amount). Gravity acting on the weight of cold lithosphere already subducted into the hotter, less dense asthenosphere pulls the plate down, causing it to bend downward. This bending, or flexing, of lithosphere is responsible for the deep-sea trench (Figure 22), with its gentle slope on the seaward side, and a steeper slope where the other plate over-rides it.

As the plate bends down to form the trench, the lithosphere seaward of the trench is flexed upwards slightly. An analogous phenomenon can be observed by taking a sheet of paper, holding one end on the top of a table and letting the other end hang over the edge. Like the weight of subducted lithosphere, the weight of the paper causes it to bend down over the edge of the table, but it also bulges slightly upward between your hand and the edge of the table. The height and width of the bulge will depend on the thickness of the sheet of paper, with lower, but wider bulges for stiffer, thicker paper. The same applies to the outer topographic rise; it will be lower but wider for thicker lithosphere. The height of this bulge in the paper can be altered by pushing horizontally or vertically on the portion of the sheet of paper that overhangs the table, and again similar processes can act on the underthrust slab of lithosphere.

Because of the thickness of the lithosphere, its bending causes another effect: a stretching of its upper surface. This stretching of the upper portion of the lithosphere manifests itself as earthquakes and normal faulting, the style of faulting that occurs when a region extends horizontally (Figure 4). Such earthquakes commonly occur after great earthquakes, as William Stauder of Saint Louis University and Lynn Sykes noted for aftershocks of the 19 March 1964 earthquake in Alaska. By contrast with the stretching of the top surface of the lithosphere, less frequent deeper earthquakes also show that at depths of 30–50 km beneath the trench, the lithosphere is shortened, or compressed, horizontally and perpendicular to the trench (Figure 24). The combination of shallow earthquakes showing a stretching of the top of the lithosphere and horizontal compression of the deeper part obviously attests to bending of the plate.

Having been bent down at the trench, the lithosphere then slides beneath the overriding lithospheric plate. Fault plane solutions of shallow focus earthquakes (see Chapter 3) provide the most direct evidence for this underthrusting. They indicate slip on a gently

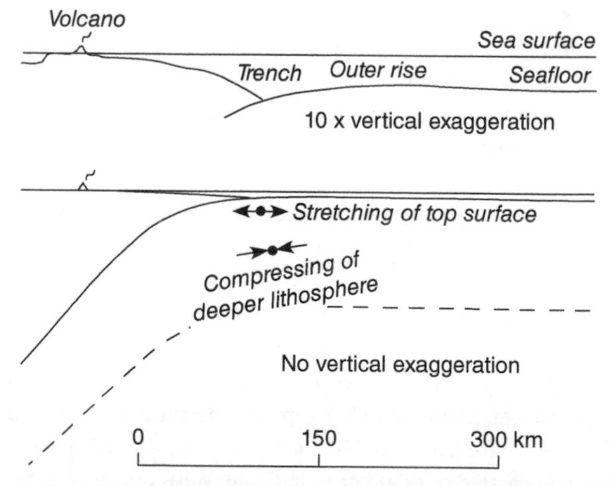

24. Cartoon cross sections of a typical deep-sea trench, outer rise, and volcanic island (top), and of flexed lithosphere (bottom) with an interpretation of fault plane solutions of earthquakes beneath the trench.

dipping plane that coincides with the distribution of earthquakes. The direction that the underthrusting plate moves with respect to the overriding plate is commonly perpendicular to the trench and to the line of volcanoes. In some regions, however, convergence between the plates is oblique to the arc; a good example is along the western Aleutians.

In great earthquakes, such as those in Sumatra in 2004, in Chile in 2010, and in Japan in 2011, the deformation of the surface of the Earth that occurs during such earthquakes corroborates the evidence for underthrusting of the oceanic lithosphere beneath the landward side of the trench. The 1964 Alaskan earthquake provided the first clear example. Shortly after the event, George Plafker, a young geologist working for the US Geological Survey, paddled in a kayak from island to island, mapping emerged and subsided shorelines. From their distribution, he inferred an underthrusting of the Pacific Ocean floor along a gently dipping plane beneath the Alaskan Peninsula. Subsequent resurveying of benchmarks to obtain horizontal changes in relative positions showed that the ocean floor had underthrust the Alaskan Peninsula by as much as 20 m, amounts comparable to those in Sumatra in 2004 and in Japan in 2011.

When Plafker did his work many geologists thought that this earthquake, and the style of faulting associated with it, might have been exceptional. One such geologist, Clarence Allen of Caltech, questioned Plafker's interpretation, but urged him to learn if similar faulting characterized the Chilean earthquake of 1960, recorded history's greatest earthquake. Largely as a result of this gentlemen's bet, and with help from Allen to fund his fieldwork, Plafker took leave from his job as a geologist in the US Geological Survey and carried out fieldwork in Chile. With no surprise to him, he and Jim Savage, also of the US Geological Survey, showed that similar thrust faulting on a gently dipping plane had occurred.

The downgoing slab of lithosphere

Although the word 'seismology' is derived from the Greek words for earthquakes and their study, many seismologists use the waves sent by earthquakes to study the Earth's interior, with little regard for earthquakes themselves. The speeds at which these waves propagate and the rate at which the waves die out, or attenuate, have provided much of the data used to infer the Earth's internal structure. The inference of a downgoing slab of lithosphere is a good example.

Because the lithosphere is much colder than the asthenosphere, when a plate of lithosphere plunges into the asthenosphere at rates of tens to more than a hundred millimetres per year, it remains colder than the asthenosphere for tens of millions of years. In the asthenosphere, temperatures approach those at which some minerals in the rock can melt. Because seismic waves travel more slowly and attenuate (lose energy) more rapidly in hot, and especially in partially molten, rock than they do in colder rock, the asthenosphere is not only a zone of weakness, but also characterized by low speeds and high attenuation of seismic waves.

Just as the bass tones, not the high-pitched voices of sopranos, from the radio of a car with all windows open in summer dominate the sound we hear, seismographs record a dominance of the lower frequencies. Waves traversing the asthenosphere are recorded more like bass than soprano voices. Waves passing through a high-speed, low-attenuation zone, however, will arrive sooner, with higher characteristic frequencies, and with larger amplitudes than those traversing normal asthenosphere. Studies of seismic waves from deep-focus earthquakes recorded at nearby stations exhibit a marked variation in their average speeds, frequency content, and amplitudes at various stations (Figure 23).

The more pronounced of the effects and the first to have been observed is the difference in attenuation of seismic waves, which we measure with a quantity called Q. High Q indicates low attenuation of P and S waves, and low Q implies that only waves with long periods (basses not sopranos) are recorded. These differences exist for P waves, the first-arriving signals from earthquakes, but are much clearer for S waves, which can propagate only through material with strength, not through liquids or gases. In general, S waves from earthquakes are poorly recorded by seismographs that are tuned to record high frequencies—the sopranos in the chorus who sing when an earthquake occurs. The amplitudes of high-frequency S waves are small, so that the predominant periods of S waves are long; commonly seismographs record only the basses of the S-wave chorus. S waves especially, but also P waves, lose much of their energy while passing through the asthenosphere. The lithosphere, however, transmits P and S waves with only modest loss of energy.

This difference is apparent in the extent to which small earthquakes can be felt. In regions like the western United States or in Greece and Italy, the lithosphere is thin, and the asthenosphere reaches up to shallow depths. As a result earthquakes, especially small ones, are felt over relatively small areas. By contrast, in the eastern United States or in Eastern Europe, small earthquakes can be felt at large distances. Shaking from the 1886 Charleston, South Carolina earthquake was felt in Boston, and modest-size earthquakes in Romania have been felt thousands of kilometres away in Moscow. Earthquakes with comparable magnitudes in California or Greece would not be felt outside their boundaries.

In 1964, fresh with a PhD in seismology, Bryan Isacks installed a seismograph network on the Tongan and Fiji Islands, the world's most active region for deep earthquakes. You can imagine the surprise when he and Jack Oliver, both then at Lamont Geological Observatory, put a seismograph on one of the Tonga Islands in the southwest Pacific and immediately noticed that the deep-focus earthquakes 1000 km from the seismographs were recorded with atypically large amplitudes. Seismograms resembled those for which paths were confined to the lithosphere, but these earthquakes occurred below the depth range not only of the lithosphere, but also of the asthenosphere (Figure 23). Stations on the nearby Fiji Islands recorded S waves with amplitudes roughly ten times smaller than those at Tonga. The paths to Tonga from local deep earthquakes follow close to the planar zone of earthquakes that dips westward beneath the Tongan Islands, whereas those to the Fiji Islands pass through typical asthenosphere (Figure 23). Oliver and Isacks reasoned that the lithosphere had plunged into the asthenosphere and provided a high-Q window through the low-Q asthenosphere that allowed S waves to travel with little loss of energy.

Since the inhabitation of the Japanese islands, the Japanese population had literally felt a similar window of low attenuation through the highly attenuating asthenosphere. Deep earthquakes occur several hundred kilometres west of Japan, but they are felt with greater intensity and can be more destructive in eastern than western Japan (Figure 25). This observation, of course, puzzled Japanese seismologists when they first discovered deep focus earthquakes; usually people close to the epicentre (the point directly over the earthquake) feel stronger shaking than people farther from it.

Independently of Oliver and Isacks, Tokuji Utsu, then at Hokkaido University, explained this greater intensity of shaking along the more distant, eastern side of the islands than on the closer, western side by appealing to a window of low attenuation parallel to the earthquake zone and plunging through the asthenosphere beneath Japan and the Sea of Japan to its west. Paths to eastern Japan travelled efficiently through that window, the subducted slab of lithosphere, whereas those to western Japan passed through the asthenosphere and were attenuated strongly.

P and S waves also propagate through the cold slab of underthrust lithosphere with higher speeds than those that pass through the surrounding asthenosphere. Consequently, waves passing through the slab arrive at seismographic stations earlier than

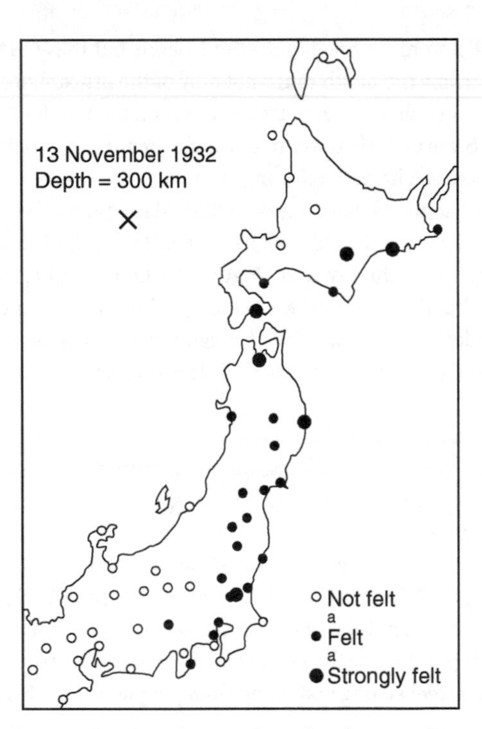

25. Map of Japan showing where and to what degree a deep earthquake west of Japan was felt. Note that it was felt more strongly in eastern than western Japan.

expected. One of the first hints of a high-speed zone came from attempts to determine the locations of earthquakes and explosions. A particularly good example of this was provided by data from the Longshot explosion, a nuclear explosion detonated on 2 October 1965 on Amchitka Island in the Aleutians and recorded throughout the world.

In determining locations of earthquakes and explosions, one compares the times at which P waves arrive at seismograph stations around the world with those calculated for appropriate distances from the source. The computer, once a person who made calculations, but by the 1960s an electronic computer, then seeks a location and time of origin that minimizes the differences between measured and calculated arrival times. Since the computer does not know where the explosion occurred, or that waves travelling down a subducted lithospheric slab have been speeded up and arrive early, it assumes that the source, the explosion in this case, must be closer to these stations than it actually is. Using data from the Longshot explosion, Lynn Sykes's computer determined a remarkably precise estimate for its location; the estimated error in the location was only 3 km. Sykes's computer's calculated location, however, was 20 km north of the actual detonation site. The high P-wave speeds in the slab dipping northward beneath the Aleutian Islands caused a mislocation much greater than the consistency of the data suggested.

Perhaps the most convincing demonstration of high wave speeds in the slab came from further study of the Tonga region, by Walter Mitronovas, then a graduate student at the Lamont Geological Observatory, working with Bryan Isacks. They located deep earthquakes in that region, and then studied arrival times of P and S waves at nearby stations. Although calculated and measured arrival times to stations on the island of Fiji agreed with one another, P and S waves travelling along the inclined seismic zone, and therefore up the slab, to seismographs on the Tongan Islands arrived exceptionally early: 5–6 seconds for P waves and 10–12 seconds for S waves. These early arrivals correspond to average wave speeds approximately 5 per cent higher than are normal for the depth range traversed by these waves. Such a large difference in speed is most easily understood to result from transmission through material with much lower temperature than normal, and therefore through subducted lithosphere.

Deep earthquakes

Zones of intermediate- and deep-focus earthquakes, which dip beneath the landward margins of 'island arc structures', appear to be about 10–15 km thick. Thus the earthquakes occur within a zone that is much thinner than either its length along the island arc or its depth range. Moreover, below a depth of about 100 km, this relatively thin earthquake zone commonly is planar in shape, though in a few regions it is warped or contorted.

These approximately planar zones of earthquakes dip into the deeper mantle at about 45° on average, but each zone is different its own way, and zones dipping as gently as 15° to as steeply as 90° can be found (Figure 26). In a number of subduction zones, earthquakes occur nearly as deep as 700 km, but in others none occurs deeper than 300 km. Moreover, the level of seismic activity in most such belts varies with depth such that where earthquakes occur at depths greater than 300 km, a minimum, if not a gap, in earthquake activity occurs between depths of 300 and 500 km. These different depth distributions seem to fall into patterns, discussed further below.

First, however, the occurrence of intermediate- and deep-focus earthquakes poses a puzzle. Shallow earthquakes occur because stress on a fault surface exceeds the resistance to slip that friction imposes. When two objects are forced to slide past one another, and friction opposes the force that pushes one past the other, the frictional resistance can be increased by pressing the two objects together more forcefully. Many of us experience this when we put sandbags in the trunks (or boots) of our cars in winter to give the tyres greater traction on slippery roads. The same applies to faults in the Earth's crust. As the pressure increases with increasing depth in the Earth, frictional resistance to slip on faults should increase. For depths greater than a few tens of kilometres, the high pressure should press the two sides of a fault together so tightly that slip cannot occur. Thus, in theory, deep-focus earthquakes ought not to occur. As early as the late 1920s, however, Kiyoo Wadati of the Japan Meteorological Agency (the government agency that studies earthquakes in Japan) had found that in and around Japan, earthquakes not

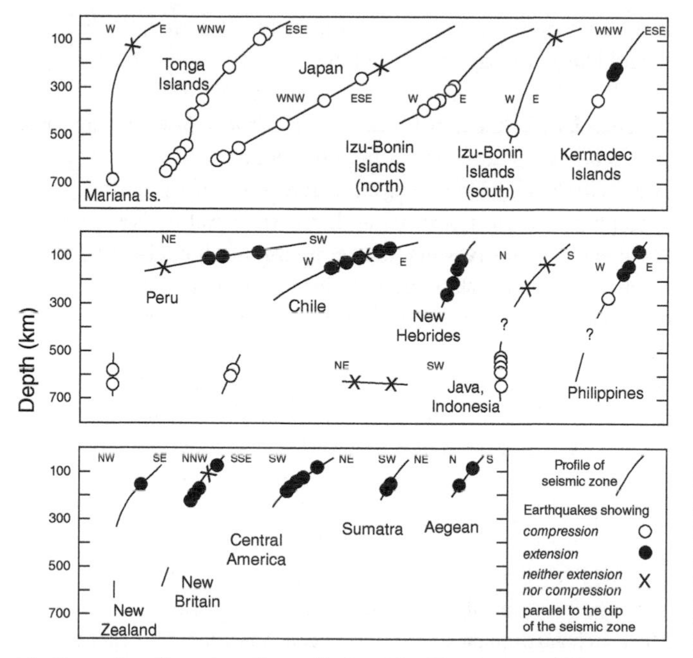

26. Examples of cross sections of seismicity (lines) across different subduction zones. Open and closed circles and X's denote earthquakes for which the P axis, the T axis, or neither of them plunges down the dip of the seismic zone.

only occurred at depths as great as 600 km, but also that they defined a planar zone dipping west beneath Japan and the Sea of Japan as far as Korea.

Within the following fifteen years, Gutenberg and Richter showed that such inclined seismic zones were typical features associated with island arcs across the Earth. Nevertheless, continued laboratory studies of rock deformed under high pressure did not reveal slip on fractures, resisted by friction, at pressures greater than roughly 10,000 times atmospheric pressure, corresponding to depths in the Earth of about 30 km. Instead, when rock was put under such pressure, rather than fracturing, it flowed, albeit slowly, like extremely viscous honey.

In general, rock, like taffy (or toffee), is brittle at low temperatures but becomes soft and flows at high temperature. The intermediate- and deep-focus earthquakes occur within the lithosphere, where at a given depth, the temperature is atypically low. Thus, it seems reasonable to suppose that the relatively hot asthenosphere is less susceptible to earthquakes than is lithosphere that has recently plunged into it. This fact allows a step to be taken toward appreciating why deep earthquakes might occur where they do, and where they do not occur, but it does not explain how deep earthquakes occur. Accordingly, the explanation for their occurrence remains a subject of discussion. Deep and shallow earthquakes do not seem to differ from one another in any measurable way, except for the prejudice that deep-focus earthquakes ought not to occur at all. Anyhow, the existence of intermediate- or deep-focus earthquakes is usually cited as evidence for atypically cold material at asthenospheric depths. Most such earthquakes, therefore, occur in oceanic lithosphere that has been subducted within the last 10–20 million years, sufficiently recently that it has not heated up enough to become soft and weak, like warm taffy.

The inference that the intermediate- and deep-focus earthquakes occur within the lithosphere and not along its top edge remains poorly appreciated among Earth scientists. In part this results from the reasonable suggestion made in 1949 by Hugo Benioff, a pioneer in seismic instrumentation at Caltech, that the inclined seismic zones found by Wadati and by Gutenberg and Richter define deep mega-faults that penetrate far into the mantle. When subduction and the plunging of lithosphere to such depths became accepted, the realization that intermediate- and deep-focus earthquakes occur within the slab went unnoticed by many non-seismologists. Now some call the deep inclined seismic zones 'Benioff zones', or more justly 'Wadati–Benioff zones', but often without the realization that the intermediate- and deep-focus earthquakes do not define a major fault in the mantle.

Why do we think that the earthquakes occur within subducted slabs of lithosphere? Suppose that the intermediate- and deep-focus earthquakes did occur on a major fault that defined the boundary between the downgoing slab and the surrounding asthenosphere. In that case, we would expect that the fault plane solutions of earthquakes (recall discussion in Chapter 3) would show one nodal plane, one possible fault plane, parallel to the seismic zone, and the other perpendicular to it. Fault plane solutions for intermediate- and deep-focus earthquakes, however, are as different as possible from that pattern. Instead, for most such earthquakes, the two nodal planes, the two possible fault planes, intersect the plane defined by the inclined seismic zone at about 45°.

When Sykes used fault plane solutions to show that transform faulting occurred (see Chapter 3), he had every reason to expect that one plane would be vertical and would trend parallel to the fracture zone along which the earthquake occurred. For intermediate- and deep-focus earthquakes, however, we cannot predict easily which of two nodal planes might be the fault plane. (In Figure 20, if the earthquake ruptured the plane perpendicular to that shown, the P-wave first motions would be identical to those shown.) In fact, this ambiguity is not important for intermediate- and deep-focus earthquakes; the orientations of the nodal planes are not the key facts. We can treat the fault plane solution as showing how the volume of rock around the earthquake source deforms, without specifying the plane that ruptured in the earthquake. Regardless of which of the nodal planes ruptures during an earthquake, we determine four quadrants surrounding the earthquake source such that two opposite quadrants move toward one another, and the other pair of opposing quadrants move apart (Figure 20). We then define the axis through the middle of the converging quadrants as the *P* axis (originally for 'pressure', but better understood as the axis of maximum com*P*ression), and the axis through the middle of the diverging quadrants as the *T* axis (originally for 'tension', but better understood as the axis of maximum ex*T*ension). For most intermediate- and deep-focus earthquakes, one of the *P* or *T* axes is aligned parallel to the dip of the seismic zone and the other is perpendicular to that zone.

When a board is deformed, deformation usually occurs such that two of the three principal stress axes lie in the plane of the board and the third is perpendicular to it. Assuming that the *P* and *T* axes define the orientations of the maximum compressive stress and maximum extensional stress, then, as I discuss below, the fault plane solutions suggest that the state of stress in the downgoing slab is what one would expect if the slab deformed like a board, or slab of wood. Accordingly, we infer that the earthquakes occurring within the downgoing slab of lithosphere result from stress within the slab, not from movement of the slab past the surrounding asthenosphere. Because the lithosphere is much stronger than the surrounding asthenosphere, it can support much higher stresses than the asthenosphere can.

Forces acting on the downgoing slab

If the intermediate- and deep-focus earthquakes resulted from stress within the downgoing plate, then the orientations of these stresses should provide information about the sources of stress, and therefore about the forces controlling the dynamics of the downgoing slab. Recall that the slab is cold, and denser than the surrounding asthenosphere. Gravity will tend to pull it down through the less dense asthenosphere. A spring allowed to hang from a skyhook, for example, or pushed down on the floor, provides an analogy. A spring dangling from a skyhook will stretch as gravity acting on its weight pulls it down. At the opposite extreme, we might detach the spring from the skyhook and push it down on the floor, so that the entire spring is compressed. Between these cases, the spring might dangle from the skyhook, so that its upper part stretches, but if its bottom part rests on the floor, that part will be compressed. Fault plane solutions of intermediate- and deep-focus earthquakes show all of these patterns.

It turns out that for nearly all deep-focus earthquakes (depths > 300 km) that have been studied, *P* axes plunge down the dip of the inclined seismic zone (Figure 26). By analogy with the spring detached from the skyhook, this suggests that the downgoing slab of lithosphere is being compressed as it plunges deeply through the asthenosphere (Figure 27). Its motion is presumably being resisted by stronger and/or denser material at greater depth. (Although temperature increases with depth, the minerals of the deeper mantle are both stronger and denser than those in the asthenosphere.) In the analogy with the spring, the floor supports some, if not all, of the weight of the spring, which is compressed.

For many years, this increased resistance to the sinking of the downgoing slab was thought to indicate that the subducting lithosphere did not penetrate into the deep mantle below 700 km. Now most Earth scientists imagine that although the slab encounters more resistance below approximately 300–400 km than above that depth, it continues to descend to greater depths.

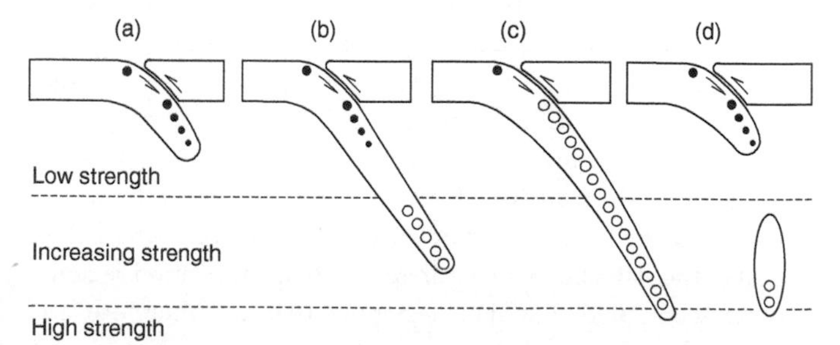

27. Simple interpretation of earthquake fault plane solutions in Figure 26. In all, a slab of lithosphere plunges beneath an over-riding plate of lithosphere. Open or closed circles indicate earthquakes for which the P or T axis plunges down the dip of the slab of lithosphere.

For most intermediate-depth earthquakes (depths between 70 and 300 km) in regions where there are no deep earthquakes at all (like Central America, Sumatra, or the Aegean), the *T* axes plunge parallel to the dip of the seismic zone (Figure 26). In this case, gravity acting on the excess weight of the slab (or of our spring hung from the skyhook) pulls the upper part of the slab (or spring) down, causing it to stretch (Figure 27).

In regions where there are both intermediate and deep-focus earthquakes, commonly the *T* axes plunge down the dip at intermediate depths, and the *P* axes of the deep earthquakes plunge down the dip of the inclined seismic zone (like beneath the Kermadec Islands or Chile in Figure 26). By analogy with the spring, the upper part of the slab is stretched by gravity pulling it down, but the increased strength or density near 700 km (the floor) supports part of the weight of the slab (or spring).

In a few regions, like beneath the Tonga Islands (Figure 26), *P* axes plunge down the dip at all depths. We presume that the slab is being forced down faster than gravity can pull it down, as with a spring detached from the skyhook and pushed down against the floor.

These observations are consistent with a cold, heavy slab sinking into the asthenosphere and being pulled downward by gravity acting on it, but then encountering resistance at depths of 500–700 km despite the pull of gravity acting on the excess mass of the slab. Where both intermediate and deep-focus earthquakes occur, a gap, or a minimum, in earthquake activity near a depth of 300 km marks the transition between the upper part of the slab stretched by gravity pulling it down and the lower part where the weight of the slab above it compresses it. In the transition region between them, there would be negligible stress and, therefore, no or few earthquakes.

An alternative explanation, however, can account for a gap in earthquake activity (Figure 27). Suppose that a deep portion of lithosphere broke free from the upper part and sank into the asthenosphere. The upper portion would behave as if there were only a short slab of lithosphere beneath it (a spring hanging from the skyhook), and the lower portion would behave as if only a short slab lay above it (a spring resting on the floor). Although this explanation may seem somewhat contrived, it appears that in some regions a deep slab has become detached from its shallower part. The distribution of earthquakes associated with subduction zones beneath Peru, the New Hebrides Islands, and New Zealand (Figure 26) shows very pronounced gaps in activity most simply explained by detached slabs, or 'slab break-off'.

Volcanoes

For some, the 'ring of fire' of volcanoes surrounding the Pacific Ocean, like Mount Fuji in Japan or Mount St Helens in the US state of Washington, as well as other notorious volcanoes elsewhere in the world like Vesuvius, Etna, and Santorini, comprise the most obvious, if not most dangerous, manifestations of subduction. Their existence in 'island arcs' once offered another puzzle, but now they can be understood as sensible products of plate tectonics, and specifically subduction.

Volcanoes occur where rock melts, and where that molten rock can rise to the surface. Obviously, rock must be hot for it to melt, and as a result, the presence of volcanoes above regions where cold slabs of lithosphere have been underthrust might seem puzzling. In winter we do not thrust ice beneath our beds to keep warm! For essentially all minerals, however, melting temperatures also depend on the extent to which the minerals have been contaminated by impurities. Those of us living in cold climates know that salt spread onto roads lowers the melting temperature of ice, so that ice turns to liquid water when air temperatures remain below 0°C. Similarly, hydrogen, when it enters most crystal lattices, lowers the melting temperature of the mineral. Hydrogen is most obviously present in water (H_2O), but is hardly a major constituent of the oxygen-, silicon-, magnesium-, and iron-rich mantle.

The top of the downgoing slab of lithosphere includes fractured crust and sediment deposited atop it. Oceanic crust has been stewing in seawater for tens of millions of years, so that its cracks have become full either of liquid water or of minerals to which water molecules have become loosely bound. Moreover, sediment covers that crust and also buries water in the interstices between sediment grains. It follows that the downgoing slab acts like a caravan of camels carrying water downward into an upper mantle desert. When the water is freed, even from the relatively cold basalt, and then penetrates into the overlying wedge of mantle above the downgoing slab, it then lowers the melting temperature of that material (like salt in ice). A relatively small amount of melt, less dense that the surrounding solid rock, then rises into somewhat warmer mantle, and enhances its tendency to melt. The resulting molten rock then makes its way through the cooler lithosphere into the crust. Some such melt cools relatively quickly in the crust, and freezes there to make granite, and some erupts at the surface to make volcanoes. Thus, plate tectonics, with the help of water, acts as a giant refinery, distilling the mantle of its materials that are light and that melt at low temperatures, and adding them to the surface scum, the continental crust.

Confirmation that sediment and crust have been subducted to depths of 100–150 km comes from an element, beryllium, that until recently was not used much, even by geologists. Some isotopes (different versions of the same element but with differing numbers of neutrons in their nuclei) of certain elements are produced only by collisions with cosmic rays. A good example is radiocarbon, or ^{14}C, a carbon atom with 6 protons and 8 neutrons, whereas most carbon, ^{12}C, consists of 6 protons and 6 neutrons. Another such 'cosmogenic' nuclide is the isotope of beryllium with 4 protons and 6 neutrons, ^{10}Be, whereas most beryllium consists of 4 protons and 4 neutrons, ^{8}Be. ^{10}Be is radioactive, and in a collection of ^{10}Be atoms, half of them decay to form an isotope of boron, ^{10}B, after 1.4 million years have elapsed (a half-life of 1.4 million years). ^{10}Be is produced only in the atmosphere and at the Earth's surface, from collisions with cosmic rays. Yet, lavas erupting at volcanoes at island arcs contain ^{10}Be.

The path for ^{10}Be, produced at or above the Earth's surface and erupted at island arc volcanoes, must begin either in the atmosphere or on the surface, then make its way to the bottom of the ocean as sediment, be transported by the ocean floor to a trench, and then be subducted with the oceanic lithosphere to depths of 130–150 km, before becoming part of the molten rock that then makes its way to the Earth's surface as lava. Moreover, it must take this path rapidly. Even if subduction occurs as rapidly as 100 km per million years, 2 million years will have elapsed before the ^{10}Be atoms will have plunged a distance of 200 km at an angle of 45° to a depth of 140 km, by which time most of the ^{10}Be atoms will have decayed. At lower rates of subduction, even fewer atoms will have avoided radioactive decay. Then, these atoms must make their way to the surface in the molten rock, so that enough of them survive to be measured.

Geochemists are a clever lot, and they have devised ways of measuring tiny fractions of rare elements. Their analysis of ^{10}Be in lavas at island arcs demonstrates that sediment and oceanic crust subducted at island arcs contribute to the volcanoes that erupt along such arcs.

The 'ring of fire' of volcanoes surrounding the Pacific Ocean, as well as other notorious volcanoes like Vesuvius, Etna, and Santorini, the deepest trenches in the seafloor, the Earth's greatest earthquakes, and the Earth's deep earthquakes all owe their existence to subduction—the underthrusting of oceanic lithosphere into the deeper mantle. Because of the lithosphere's strength, before it plunges into the asthenosphere, it bends down gradually to form deep-sea trenches in the ocean floor, and also flexes up slightly to form an outer topographic rise (Figure 22). The lithosphere does not slide smoothly beneath the overriding plate of lithosphere but moves in lurches during earthquakes, sometimes as much as 20 m or more in the world's greatest earthquakes. The downgoing slab of lithosphere carries water in cracks in oceanic crust and in the interstices among sediment grains, and when released to the mantle above it, hydrogen dissolved in crystal lattices lowers the melting temperature of that rock enough that some of it melts. Many of the world's great volcanoes, like Fuji in Japan, Mount St Helens in the western USA, or Vesuvius and Etna in Italy, begin as small amounts of melt above the subducted slabs of lithosphere. The cold, strong slabs of lithosphere that plunge into and penetrate through the asthenosphere not only provide windows through it by allowing seismic waves to propagate with only modest loss of energy, but also host all of the world's deepest earthquakes. When they proposed that oceanic lithosphere plunged beneath the Tongan Islands, Oliver and Isacks could not have imagined all of the phenomena that this process could explain.

Chapter 5
Rigid plates of lithosphere

Plate tectonics follows the maxim attributed to Einstein: 'Everything should be made as simple as possible, but no simpler.' Plate tectonics brings simplicity to the Earth because (in most regions) plates of lithosphere behave as rigid, and therefore undeformable, objects. The high strength of intact lithosphere, stronger than either the asthenosphere below it or the material along the boundaries of plates, allows the lithospheric plates to move with respect to one another without deforming (much).

Analogies can be easily made with ice floes moving with respect to one another over the ocean, or with ships at sea. If we know how the bow, stern, and mast of a ship are moving with respect to a harbour, then we know how the entire ship is moving. The same does not apply to the water; the movement of the water varies over short distances, as a ship moves through it, as wind blows over it, as the moon and sun exert forces that create tides, etc. By most interpretations of the word 'simple', the movement of a ship over the seafloor is much simpler than the movements of the surrounding water.

Dan McKenzie of Cambridge University, one of the scientists to present the idea of rigid plates, often argued that plate tectonics was easy to accept because the kinematics, the description of relative movements of plates, could be separated from the dynamics, the system of forces that causes plates to move with respect to one another in the directions and at the speeds that they do. Making such a separation is impossible for the flow of most fluids, like water in the ocean or the air in the atmosphere, whose movement cannot be predicted without an understanding of the forces acting on separate parcels of fluid.

In part because of its simplicity, plate tectonics passed from being a hypothesis to an accepted theory in a short time. In my case, that interval of time was measured in hours, the time it took to read Jason Morgan's paper presenting the idea in the autumn of 1967, eight months before that paper was published. By the time that Dan McKenzie arrived to be a post-doctoral fellow at Lamont Geological Observatory (of Columbia University, where I was a graduate student) in December 1967, and shortly before his paper that independently presented the idea of rigid plates was published, my PhD advisers and fellow students were not debating the basic idea, but moving forward to address its implications and the reasons that the Earth obeyed the rules of plate tectonics.

Rigid plate motion

As a sphere, the Earth brings some simplicity to the description of relative motion of plates. If we describe the relative motion of two blocks of wood, like those children play with, or of a piece of furniture in a room, as we move it across the floor, there are two parts to the movement. As we move a table across the floor, we 'translate' it, by giving it a velocity, the speed and the direction of movement, until we bring it to a halt again in a new place (Figure 28). We can also rotate the table as we move it, and that rotation introduces another aspect to the description of its movement. Moreover, the 'translation' of the table (the distance and direction that we move it) and its rotation can be independent of one another, due to different forces applied to the table and different interactions of the table with its surroundings.

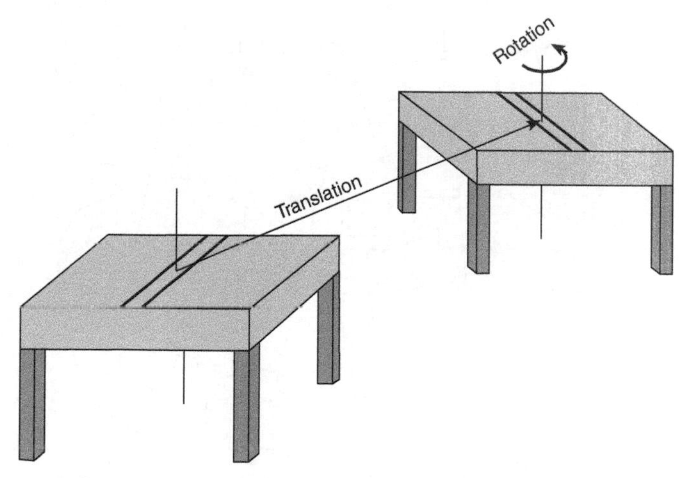

28. Relative motion of objects on a flat surface; 'translation' from one place to another and rotation about an axis through its centre.

By contrast, for plates that move over the surface of a sphere, all relative motion can be described simply as a rotation about an axis that passes through the centre of the sphere. The Earth itself obviously rotates around an axis through the North and South Poles. Similarly, the relative displacement of two plates with respect to one another can be described as a rotation of one plate with respect to the other about an axis, or 'pole', of rotation (Figure 29). For example, we describe the movement of the North America plate, which includes most of the North American continent, with respect to the Eurasia plate, which includes most of Europe and northern Asia, by a rotation about an axis that pierces the Earth's surface in eastern Siberia (Figure 30) (and 180° from it, south of South America), or more precisely by an angular velocity.

Burgeoning scientific fields commonly appeal to terminology that is easily understood by all, so as to enable those unfamiliar with the ideas to grasp them. Accordingly, such axes of rotation were

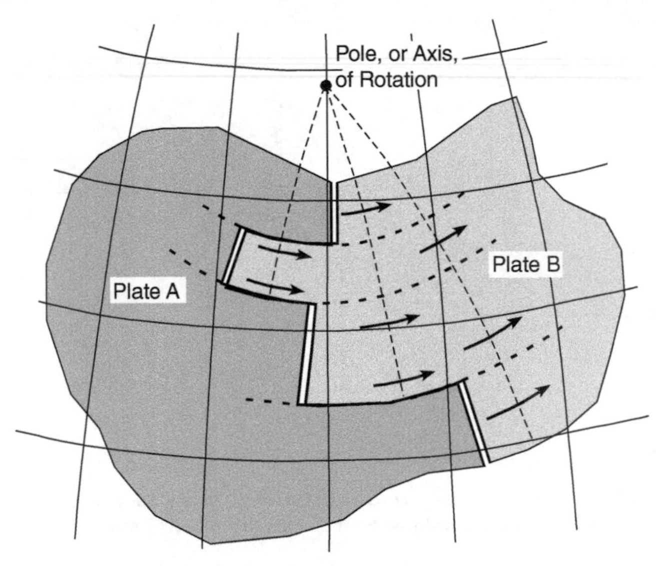

29. Relative motion of plates on a sphere about a pole of rotation. Thin solid lines define lines of latitude and longitude. Two plates, A and B, spread apart at segments of mid-ocean ridge (marked by parallel lines) and slide past one another at transform faults (single arcuate lines), with fracture-zone continuations (dashed lines).

called poles of rotation; by analogy with the Earth's spin axis and North and South Poles, the axis of rotation describing the relative movement, or displacement, of two plates is also a pole. Practitioners in mature fields often burden their disciplines with obscure terminology that limits access, much as teenagers invent slang that their parents cannot understand. Now plate tectonics is encumbered by the misleading term, 'Euler vector'. Leonhard Euler, a 19th century Swiss mathematical physicist, showed that movement on the surface of a sphere can be described by a rotation, and physicists describe rotations in terms of 'Euler angles', which, however, share little with 'Euler vectors'. It pains me to realize that plate tectonics has become so esoteric as to require obfuscating terminology to isolate those in-the-know from curious outsiders.

30. Map of the Atlantic and surrounding regions showing plate boundaries (solid lines) and axes, or poles, of rotation about which pairs of plates rotate with respect to one another.

The description of the motion of one plate with respect to another, relative plate motion, as a rotation about an axis is merely a mathematical artifice. A search in Siberia for the axis of rotation describing the relative motion of North America and Eurasia is reminiscent of the search for a pot of gold at the end of a rainbow. Unlike going to the end of a rainbow, one can pinpoint the axis about which North America and Eurasia rotate relative to one another, but like the rainbow, there is no pot of gold, or anything else worthy of a visit to the point where the rotation axis intersects the Earth's surface.

If we know how two cars are moving on highways, we can calculate how they move with respect to each other (whether they will move farther from one another, whether they will collide, etc.). Similarly, if we know how two plates, for example Eurasia and Africa, move with respect to a third plate, like North America, we can calculate how those two plates (Eurasia and Africa) move with respect to each other. A rotation about an axis in the Arctic Ocean describes the movement of the Africa plate, with respect to the North America plate (Figure 30). Combining the relative motion of Africa with respect to North America with the relative motion of North America with respect to Eurasia allows us to calculate that the African continent moves toward Eurasia by a rotation about an axis that lies west of northern Africa. Recall that in the 1960s (before we had GPS measurements), we could determine rates of relative plate motion only from magnetic anomalies formed at spreading centres, not where two plates converge and one is destroyed. By combining the known relative motion of pairs of plates, however, we can calculate how fast plates converge with respect to one another and in what direction. Such calculations, for example, show that northwestern Africa (Morocco) moves northwest toward Spain at about 5 mm/yr, but Egypt moves northward toward Europe much faster, at ~15 mm/yr. The Mediterranean Sea is slowly becoming narrower.

Now, in the 21st century, we can measure how plates move with respect to one another using Global Positioning System (GPS) measurements of points on nearly all of the plates. Such measurements show that speeds of relative motion between some pairs of plates have changed a little bit since 2 million years ago, but in general, the GPS measurements corroborate the inferences drawn both from rates of seafloor spreading determined using magnetic anomalies and from directions of relative plate motion determined using orientations of transform faults and fault plane solutions of earthquakes. When, more than 45 years ago, plate tectonics was proposed to describe relative motions of vast terrains, most saw it as an approximation that worked well, but that surely was imperfect. As will be discussed in Chapter 6, plate tectonics is imperfect, but GPS measurements show that the plates are surprisingly rigid. Although it is hard to find a rock at the Earth's surface that did not undergo some deformation in its past, most of us did not expect plates to be as rigid as they have proven to be.

Rotations about axes through the centre of the Earth not only describe how fast and in what directions plates move with respect to one another, but they also can be used to describe where two plates lay with respect to one another at different times in the past. To fit the coasts of South America and Africa together, we may rotate one coast toward the other about an axis that penetrates the Earth's surface just north of the Azores Islands. This axis lies about 15° south of the axis near southern Greenland about which Africa and North America diverge at the present time (Figure 30).

Recall that the magnetic anomalies in the ocean form at precise times. Hence, they define lines on the Earth's surface that mark seafloor of the same age, or isochrons (from *iso* meaning 'equal' and *chronos* meaning 'time' in Greek, but referring to age here). Thus, if one wants to know where two plates lay with respect to one another in the past, one maps such isochrons and then rotates one isochron to overlie the other. By doing so, one skips past the intervening seafloor, all of which is younger than that of the isochron in question. As some magnetic anomalies are more easily recognized and identified than others, we make reconstructions for their corresponding ages.

Just as we can combine relative movements of different plates, we can combine reconstructions of relative positions of different plates. For example, we can reconstruct the relative position of the North America plate relative to the Eurasia plate at, say, 32 million years ago, and we can reconstruct the relative position of the Africa plate relative to the North America plate at the same time. Thus, we can determine where the Africa plate lay relative to the Eurasia plate at that time, and at other times. Although we cannot measure the relative positions of Africa and Eurasia directly, we can determine them from their positions relative to North America.

Tests of plate tectonics

Most theories vanish before many scientists learn of them, because they make predictions, and experiments show the predictions to be wrong. Plate tectonics too made predictions. Perhaps, now, the most definitive test is that described above using GPS measurements to corroborate both the assumption that the plates are rigid and the movements of plates with respect to one another that had been deduced from magnetic anomalies, transform faults, and fault plane solutions of earthquakes. For this, a technique imagined surely by only a trifling few in the 1960s corroborated what had been inferred from completely different data.

Tests were made in the 1960s as well. In their presentation of the idea, Dan McKenzie and Robert Parker, of Scripps Institute of Oceanography, showed that the directions of relative movement between the Pacific and North America plates matched those predicted by a rotation of one of these plates with respect to the other about an axis in eastern North America (Figure 30). They relied entirely on fault plane solutions of earthquakes (see Chapter 3), which show Baja California sliding northwest relative to the rest of Mexico, the northeast Pacific Ocean floor sliding in a more northerly direction past southeast Alaska and western Canada and diving northwestward beneath the Alaska Peninsula, and farther west, the North Pacific floor plunging obliquely northwestward beneath the Aleutian Islands, and northwestward beneath the peninsula of Kamchatka and the Kurile Islands of eastern Russia. The consistency of relative motion along a boundary whose orientation varies from NW–SE in the Gulf of California, to NNW–SSE in southeast Alaska, then NE–SW along the Alaska peninsula–Aleutian island arc and the Kamchatka–Kurile region provided a good test of plate tectonics.

Additional tests promptly followed. Recall that movement of a merry-go-round is slow at its centre and fast on the edges. If relative plate motion is described by a rotation about an axis, similarly two plates separating from one another should do so slowly near that axis (analogous with the centre of the merry-go-round) and increase to a maximum 90° from the axis (analogous with the edge of the merry-go-round). In his presentation of plate tectonics, Jason Morgan, of Princeton University, showed that spreading rates at mid-ocean ridges indeed increase smoothly with distance from poles of rotation. For example, the rate that the North America and Eurasia plates separate increases from less than 10 mm per year in the Arctic, not far from the pole of rotation in Siberia (see Figure 30), to more than 20 mm per year in the central Atlantic Ocean.

When Xavier Le Pichon heard Morgan's presentation of plate tectonics in April 1967, he dropped what he was doing and applied the idea to the whole Earth. He divided the Earth into six plates: Eurasia, Africa, India-Australia, (much of) the Pacific, North and South America, and Antarctica (Figure 6). Data for smaller plates, like those in the eastern Pacific, were too sparse to allow those areas to be included. Le Pichon had access to all of the marine magnetic anomalies and bathymetric data that Ewing's ships had been gathering over the past decade. More than others, he was prepared to calculate seafloor spreading histories for the various oceans. Thus, he could calculate directions that pairs of plates moved toward one another at subduction zones. Isacks, Oliver, and Sykes promptly compared them with directions determined from fault plane solutions of earthquakes and found agreement with Le Pichon's predictions.

Le Pichon not only calculated how these plates move with respect to one another today, but he also determined histories of relative movements of pairs of plates over the preceding 80 million years. For many geologists, the most important consequence of the recognition of plate tectonics was that it brought confirmation of continental drift. As important, plate tectonics allowed continental drift to be determined with considerable precision.

This became immediately clear to Clark Burchfiel, now at Massachusetts Institute of Technology (MIT), who in 1968 had carried out fieldwork in Yugoslavia, and returned home to discover Le Pichon's paper. By knowing the relative motions between Africa and North America and between Europe and North America, Le Pichon had predicted the motion between Africa and Europe for the previous 80 million years, a history that included not only the present-day convergence between Africa and Europe, but also periods when these two regions moved in different directions relative to one another, including a period when they diverged from one another. This history of relative plate motion wrote a complex history on the rock record of the region affected by the relative movement of Africa and Europe. Nevertheless, Burchfiel had inferred from the geologic history of this rock the times of major changes and had inferred what tectonic processes (convergence, divergence, and directions of relative movement) might be occurring at these times. Le Pichon's calculated plate motions, from data solely in the Atlantic Ocean, predicted many of Burchfiel's observations and inferences. A new idea becomes believable when it predicts something that has not yet been measured or explained, especially when the idea is really trying to explain other facts.

The geologic history of the western United States offered another test. At present, the Pacific plate west of California slides northwest past the North America plate. Slip on the San Andreas fault, host to the earthquake that destroyed much of San Francisco in 1906, accommodates roughly two-thirds of that relative movement. In the late 1960s, tens of person-years of field geology had demonstrated: (1) that the western part of California, its Coast Ranges, consisted of rock that had been on the ocean floor roughly 100–150 million years ago; (2) that the eastern part of California consisted mostly of granite that had been intruded at the same time; and (3) that slip on the San Andreas fault had offset that old rock. Most had inferred that the San Andreas fault also was an ancient feature, for the geologic record offered little to contradict that view.

At that time, the late 1960s, Tanya Atwater was a graduate student at Scripps Institution of Oceanography, and was studying the seafloor of the Pacific to the west. Using only magnetic anomalies and fracture zones in the Pacific and the rules of plate tectonics, she pointed out that ocean floor had been subducted beneath California until as recently as 30 million years ago, despite there being no obvious geologic record of that event, and that the San Andreas fault too must be a young geologic feature, one that formed after that subduction stopped (since 30 million years ago).

The discussion that ensued was rendered complex, not so much because a graduate student was telling all of the authorities that they were wrong, but because that student was an unaggressive woman. (One eminent colleague jokingly described her as politely answering questions by those who were misguided, while gently cutting them off at the knees. After she had disassembled the views of a doubter on one occasion, that colleague asked the questioner, 'You didn't feel that, did you?') Of course, some were appalled that a student who had not mapped the geology could tell those who had spent their lives doing so that her data from another part of the world showed that their interpretations of their data were wrong. Consistent with Californians often being at the vanguard, however, most caught on quickly, and her paper became an instant landmark. A closer look at the geology corroborated her inferences, and with her study, the rules and quantitative aspects of plate tectonics made their impact on the continental geology.

I had my own epiphany on a related aspect of plate tectonics many years later. For decades, geologists working in the Andes had inferred that the high mountain range had been built in relatively short phases separated by longer, more quiescent intervals. My reading of the literature, burdened by my prejudices about how the Earth worked, left me sceptical of those claims. So, Federico Pardo-Casas, a Peruvian graduate student at MIT, and I set out to reconstruct the history of subduction of seafloor beneath western South America, by combining the histories of how South America moved away from Africa, Africa from India, India from Antarctica, and so forth. Much to my surprise, we found not only that the rate of convergence between South America and the seafloor to its west had varied over the past 60 million years, but also that the intervals with faster convergence were simultaneous with the periods when geologists had inferred rapid building of the Andes. For most of us, the most convincing results are those that prove us wrong!

To sum up, the strength of the lithosphere provides the glue that bonds seafloor spreading, transform faulting, and subduction, the essence of plate tectonics. Because of that strength, essentially rigid plates of lithosphere move with respect to one another across the surface of the Earth. Their rigidity allows the descriptions of their rates of relative motion and of their total displacements to be described by rotations about axes passing through the centre of the Earth, or poles of rotation. Long histories of plate motion can be reduced to relatively few numbers, the latitudes and longitudes of the poles of rotation, and the rates or amounts of rotation about those axes.

Among tests of plate tectonics, none is more convincing than the GPS measurements of changing positions of tens to hundreds of points on most plates; their relative velocities corroborate the rates inferred from magnetic anomalies, orientations of transform faults, and fault plane solutions of earthquakes along plate boundaries. Moreover, numerous predictions of rates or directions of present-day plate motions and of large displacements of huge terrains have been confirmed many times over. As with most scientific subjects, the exceptions to plate tectonics, the regions where plate tectonics failed, became the next target of research.

Chapter 6
Tectonics of continents

'Tectonics' is a geological term that refers to large-scale processes affecting the structure of the Earth's crust, and particularly the structure that results from deformation of the crust. The essence of 'plate tectonics' is that vast regions move with respect to one another as (nearly) rigid objects. Therefore the words 'plate tectonics' provide an ironic twist: the study of 'plate tectonics' has focused largely on the study of places in which the structure of the Earth's crust does not change, where there is no tectonics at all. Although plate tectonics works well for the vast regions beneath the deep ocean, for instance beneath the Pacific Ocean and at its margins, the study of tectonics itself, of processes affecting the crust, has in fact long been largely the study of continents.

One difference between rigid plates beneath oceans and mountain ranges in continents is illustrated well by the distribution of earthquakes (Figure 3). Boundaries between plates in oceanic regions are narrow—mid-ocean ridges, transform faults, and subduction zones—in many cases consisting of only one fault. The vast regions between these belts of earthquake activity, the plates themselves, undergo little deformation—with few earthquakes—and behave as (nearly) rigid objects.

Where the boundary between two plates passes into a continent, however, earthquakes are spread over wide areas. For example, the Himalaya defines the northern edge of the effectively rigid India plate, but the southern edge of the rigid part of the Eurasia plate lies 1000–3000 km farther north, and a wide zone of earthquake and tectonic activity characterizes the region between the two effectively rigid plates (Figure 3). Although not all geologists agree, I argue that the rules of plate tectonics do not apply to such regions.

Continents differ from oceanic regions in a variety of ways, but most obviously in the thickness of the crust, the outer layer of the Earth that overlies the mantle, which contains most of the Earth's mass (Figure 1). Beneath deep oceans, the thickness of crust is only 7 km, but beneath continents, thicknesses range from as little as 25 km, or even thinner in rare places, to more than 75 km beneath very high regions, with an average of 35 to 40 km. Beneath all continents the crust is much thicker than it is beneath deep oceans. The thicker crust beneath continents than oceans affects the lithosphere in two ways that make continental lithosphere less prone to rigid-plate behaviour than oceanic lithosphere is.

Most obviously, the thick crust on the top of the mantle lithosphere makes the continental lithosphere more buoyant than oceanic lithosphere. Recall that temperatures in the lithosphere are lower than those in the asthenosphere. Therefore (in general) when lithospheric material is placed at the same depth as asthenospheric material, and hence at the same pressure (for density increases with pressure), the lithosphere is the denser. Oceanic crust, 10–15 per cent lighter than the mantle (see Chapter 1), provides only a negligibly thin buoyant top to the oceanic lithosphere. When oceanic lithosphere subducts, its crust is carried down with the mantle portion of the lithosphere. (In fact, subducted oceanic crust might transform to denser material that makes it negatively buoyant.) When continental lithosphere enters a subduction zone, however, the thick, buoyant crust resists subduction, just as a life-preserver tossed to a 'man overboard' prevents him from sinking and drowning. Rather than following its underlying mantle lithosphere deep into asthenosphere, the upper part of the overlying continental crust detaches from the lower crust and mantle lithosphere, and piles up in front of the subduction zone to build a mountain range.

The Alps of Europe and the Himalaya in Asia illustrate this phenomenon. A collision of the India and Eurasia plates created the Himalaya. As India's ancient northern margin slid beneath the southern edge of Eurasia, some of India's upper crust detached from lower crust beneath it. Then, as the lower crust of India continued to plunge beneath Eurasia's southern margin, a stack of thin slices grew to form the Himalaya. At present, intact India lithosphere underlies the southern part of the Himalaya (Figure 31), and the rock cropping out in the Himalaya consists largely of slices of India's crust. Similarly, a collision of the southern edge of Eurasia plate in Europe with a promontory on the northern edge of the Africa plate, which includes most of the Italian peninsula, created the Alps. Slices of the Eurasian crust became detached from the underlying mantle portion of the

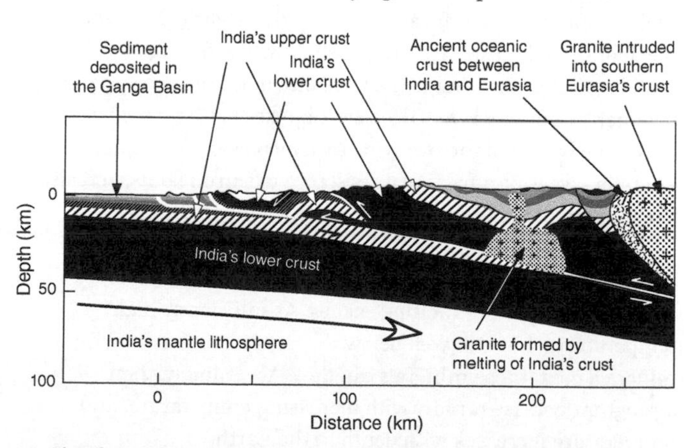

31. Simple cross section through the Himalaya. The lithosphere beneath India slides beneath Eurasia's southern margin (on the right side). Most of the rock in the Himalaya consists of slices of rock that once were the top part of India's crust.

Eurasian lithosphere that plunged beneath the Po Plain in northern Italy. The Alps were built by thin slices of Eurasia's southern edge becoming stacked atop one another.

A second difference between continental and oceanic lithosphere is arguably the more important for understanding differences between continental tectonics and plate tectonics. Minerals comprising the mantle seem to be stronger than most of those in the crust, and as a result, at the depths where oceanic lithosphere is strongest, at depths of 20–40 km, continental lithosphere is weak. To appreciate this difference, we must have an image of how strength varies with depth.

Near the surface, where temperatures are low, rock is brittle. It fractures. For brittle rock to deform, stresses must overcome the frictional resistance to slip on the faults and fractures in the Earth's crust and uppermost mantle.

We all rely on the frictional resistance of our feet in our daily lives to avoid slipping and falling, and we know that sliding an object across the floor is made easier by lifting it slightly. The frictional resistance to sliding is proportional to the force that holds the two surfaces together (see discussion of deep earthquakes in Chapter 4). In the Earth, the weight of the overlying rock increases with depth. Therefore we expect frictional resistance to slip on faults, and also the strength of the rock, to increase with depth (Figure 32).

Like butter or ice cream, all minerals weaken as temperature increases even before melting ensues. At sufficiently high temperatures, but still well below the temperature at which minerals melt, those minerals can flow. Accordingly, their strengths decrease rapidly with increasing temperature, and temperature increases with depth in the Earth.

Minerals like quartz, feldspar, and mica, which combine to make rocks of different types, deform differently when put under stress.

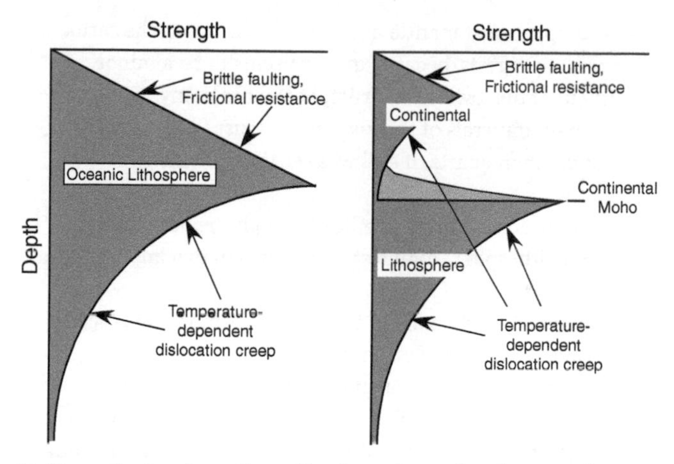

32. Brace-Goetze strength profiles from the surface downward through oceanic (left) and continental (right) lithosphere. The Moho, the boundary between crust and mantle (see Chapter 1), marks a boundary between layers with different minerals with different strengths.

For example, quartz, a major constituent of the crust, flows readily (on geological time scales) at a temperature as low as 350°C, but olivine, the main constituent of the upper mantle, remains brittle and does not flow until the temperature exceeds 600–700°C. Even in young oceanic crust, temperatures do not reach 300°C, and both the crust and the uppermost tens of kilometres of the mantle beneath oceans can be brittle. At these depths, the oceanic lithosphere deforms by slip on faults during earthquakes. Because temperature increases with depth, and because continental crust is much thicker than oceanic crust, temperatures in continental crust commonly exceed those at which quartz and other crustal minerals can flow. As a result there seems to be a layer of low strength within continental crust, at least in places where the crust is not unusually cold or thin (Figure 32). In some such regions, the uppermost mantle remains strong, and even brittle, so that the strength profile has been likened to a 'jelly (or jam) sandwich'. The contrast in strength between the lowermost crust and uppermost mantle need not be abrupt, as the cartoon in Figure 32 suggests, because quartz seems to be a minor constituent of the lowermost crust, at least in many regions, and the main minerals of the lowermost crust (such as feldspar) are stronger than quartz, if still weaker than olivine.

The differences in strength profiles through the oceanic and continental lithosphere manifest themselves in two important ways.

First, the low-strength zone in the crust in some regions seems to allow the upper crust to detach from the lower crust and uppermost mantle. The clearest examples are the thin sheets of crust, called 'nappes' in geology (the French word for sheets and tablecloths), that are stacked atop one another in some mountain ranges, like the Alps and the Himalaya (Figure 31).

Second and more important, oceanic lithosphere is, in general, much stronger than continental lithosphere; oceanic lithosphere reaches its maximum strength at depths of ~20–40 km, in the depth range where the continental lithosphere can be especially weak. Whereas the strength of oceanic lithosphere depends on the strength of olivine in the mantle at depths of 20–40 km, continental lithosphere lacks that core of high strength simply because crustal minerals, like quartz rather than olivine, occupy that depth range. That cold core of olivine at 20–40 m depths allows oceanic lithosphere to form strong plates that resist deformation, and the absence of strength in that depth range in most (but not all) continental regions facilitates their deformation over broad regions.

To the casual observer, the most obvious consequences of this weak continental lithosphere are mountain ranges. If the lateral extent of crust in some region decreases, because the surrounding lowlands move toward one another, the intervening crust must thicken. Because the crust is less dense than the mantle, the resulting thickened crust will stand higher than the surrounding lowlands, to make a mountain range. As discussed in Chapter 1, an analogy is commonly drawn with icebergs. The ice that we see above the sea is only 10 per cent of the iceberg; 90 per cent lies below sea level, hidden from view. Similarly, if we thicken crust to make a mountain range 2 km high, we must actually make the crust thicker than it had been by nearly 15 km. Most of the excess crust is stored in a 'crustal root'. The average thickness of crust beneath the high Andes and the Tibetan Plateau is roughly 70 km, some 30–35 km thicker than most continental crust, but the surface stands only 4–5 km higher than the surrounding lowlands. These are examples of Archimedes' Principle, which for the Earth we call isostasy.

Readers may be quick to see an inconsistency. Before, we have treated the lithosphere as a layer of strength, but now for continents we treat it as a deformable, if very viscous, fluid. Indeed, continental lithosphere in many places, and especially beneath wide mountain belts, is not the layer of strength that oceanic lithosphere is (Figure 32). The combination of relatively weak middle and lower crust and the absence of a strong mantle layer make the entire column of lithosphere more susceptible to deformation than is oceanic lithosphere. In the discussion of mountain ranges that follows, in some cases the strength of lithosphere plays a key role in the construction and support of mountain ranges, but in others, continental lithosphere is sufficiently weak that on geologic time scales it does behave like a fluid. Let us consider some different mountain ranges to illustrate these differences.

Central Andes

At subduction zones surrounding the northern and western Pacific Ocean, the Pacific plate plunges beneath the volcanoes of the 'island-arc structures' along a single major thrust fault. Except for the occasional volcanic eruption, the overriding plate leads a quiet existence and undergoes little activity. By contrast, the Central Andes of Peru, Bolivia, Chile, and Argentina host not only the highest peaks, but also the highest plateau, the Altiplano, found outside of eastern Asia. Oceanic lithosphere, in this case the Nazca plate (Figure 6), plunges eastward beneath the west coast of South America, and beneath the belt of volcanoes there. Unlike volcanic islands of, for example, the Aleutian or Tongan Islands, the volcanoes along the Andes have been built on high terrain. That high terrain has developed in large part because during the past few tens of millions of years the crust along the western edge of South America has been compressed horizontally and thickened.

The sustained underthrusting of oceanic lithosphere beneath the west coast of South America and the heating of the crust associated with the volcanoes weakened the western edge of the South American lithosphere. The continued subduction of the lithosphere has compressed that lithosphere horizontally, and because of isostasy, its thickened crust stands high.

Perhaps the clearest manifestation of that horizontal shortening is the wide belt of folded rock along the eastern side of the Andes. Layers of sedimentary rock have been folded, like a rug too thick to fit under a door as it is opened, and some layers have been thrust atop others in what geologists call a 'fold-and-thrust belt'. This same style of deformation built a fold-and-thrust belt along the eastern flank of the Canadian Rocky Mountains (Figure 33) and

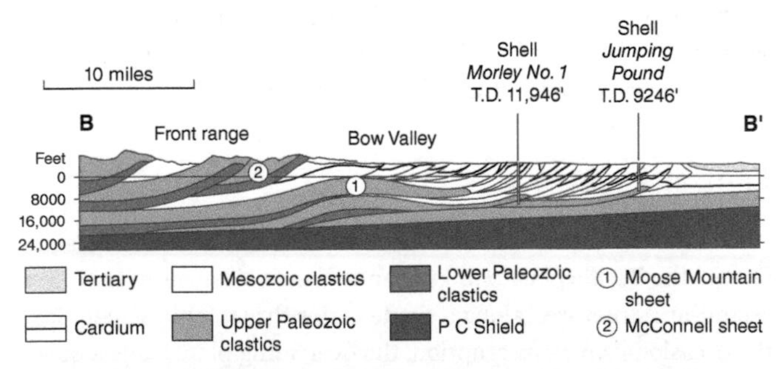

33. The fold-and-thrust belt of the Canadian Rockies. Layers of sedimentary rock, shown as dark grey, light grey, and white, have been folded and thrust atop one another, as their basement (P C Shield), shown in black, has slid westward beneath the folded and faulted sedimentary rock.

another that makes the Appalachian Mountains of Pennsylvania, Virginia, and neighbouring states. The layers in such folds that are most resistant to erosion often form ridges and therefore a 'ridge and valley' landscape; the Pennsylvania Turnpike highway uses tunnels to pass through such ridges.

Eastern Asia

The width of the Andes reaches 700 km, but mountains are being built over a region as wide as 3000 km in eastern Asia (Figure 34). Whereas the plates beneath oceanic regions are defined by belts of earthquakes that surround them and mark boundaries between plates, in continental regions, earthquakes occur widely (Figure 3). Since 1897, seventeen earthquakes comparable in magnitude to the 1906 San Francisco earthquake have occurred in eastern Asia. Only the four of them that occurred along the Himalaya might be considered to have occurred on a plate boundary. The rest ruptured faults that do not separate two plates—within the Tibetan Plateau, within the Tien Shan, or major strike-slip faults in Mongolia (Figure 34); they contribute to a pattern of ongoing geologic deformation of the Earth's crust. Some of these earthquakes took hundreds of thousands of lives, but others occurred in such remote areas that few felt them.

Essentially all of this widespread deformation and earthquake activity across eastern Asia results from India colliding with Eurasia's ancient southern edge and then penetrating deep into the Eurasian continent. The combination of the buoyancy of India's lithosphere, because of its continental crust, and a relatively weak Eurasian lithosphere has led to widespread deformation throughout southeastern Eurasia. Using the approach discussed in Chapter 5, we can calculate where India lay at different times in the past (Figure 35), for we know where the India plate lay with respect to the Africa plate, where Africa lay with respect to the North America plate, and where the North America plate lay with

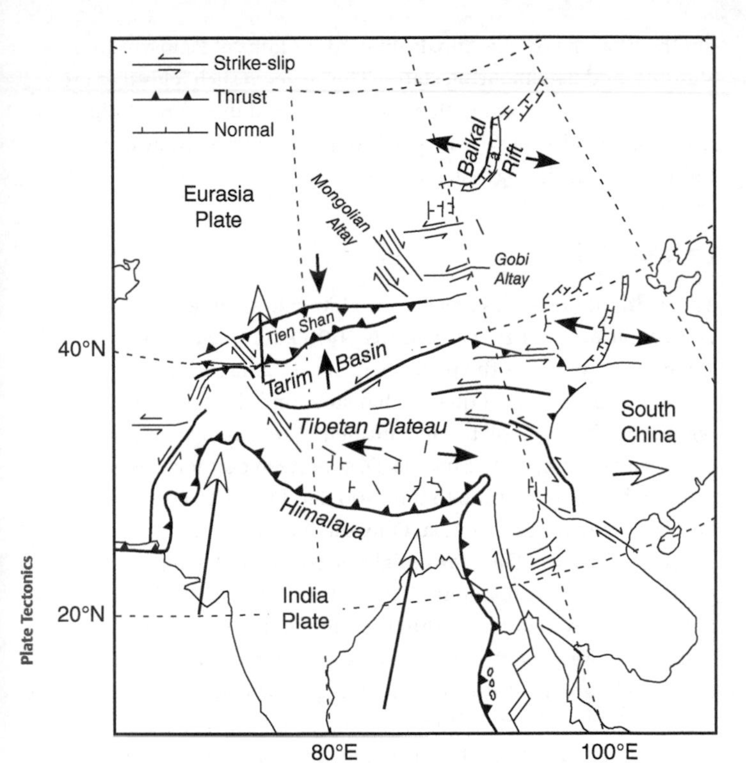

34. Map of eastern Asia showing coastlines and major active faults (black lines). Arrows with white arrowheads show the movement of large regions (India, Tarim Basin, and South China) with respect to Eurasia. Dark black arrows pointing toward or away from one another indicate convergence or divergence across the intervening regions.

respect to the Eurasia plate. Thus, we can calculate past positions of the India plate, with the Indian subcontinent as its passenger, with respect to the Eurasia plate. At the time of collision, 40 to 50 million years ago, India lay 2000–3000 km south of its present position. Although its northward march has slowed since that time, it continues to move north-northeastward toward Eurasia at 35 to 40 millimeters per year. What we cannot know well,

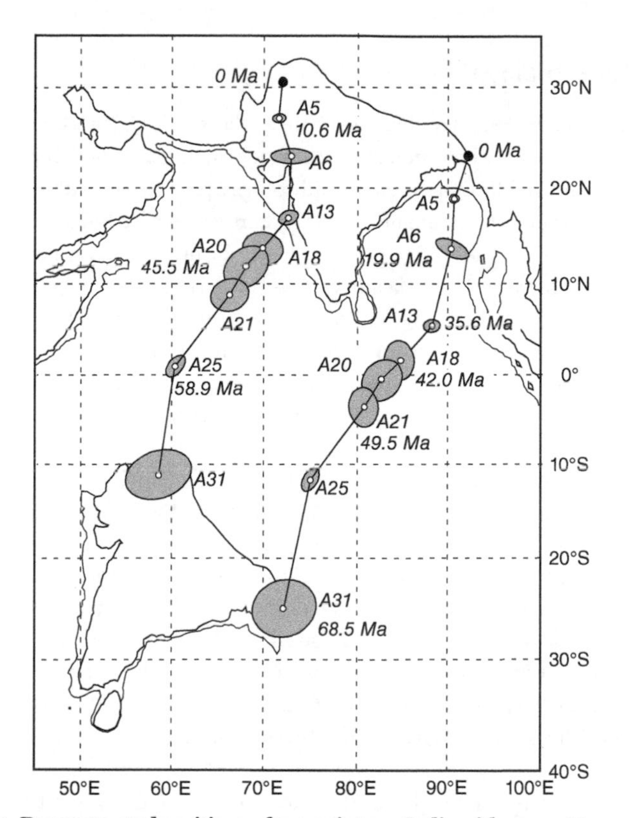

35. Reconstructed positions of two points on India with respect to Eurasia at the times of clearly defined magnetic anomalies, such as A31 (Anomaly 31) at 68.5 million years ago (Ma). Shaded ellipses show uncertainties in reconstructed positions.

unfortunately, is the northern extent of the Indian subcontinent prior to its collision with southern Eurasia 50 million years ago; most of its remnants are buried beneath southern Eurasia.

Let us examine separate parts of eastern Asia including both some aspects for which plate tectonics helps us to understand mountain ranges there and others for which plate tectonics provides little guide.

The Himalaya

As noted, the Himalaya has been built as slices of India's northern margin that have been detached from the remaining, underlying, intact Indian lithosphere, as that lithosphere plunged beneath the ancient southern margin of Eurasia (Figure 31). Before India reached the southern edge of Eurasia, oceanic lithosphere, which lay between India and Eurasia and carried India as a passenger, plunged beneath the southern edge of Eurasia, just as lithosphere of the Nazca plate plunges eastward beneath the Andes today. The Indian subcontinent followed that oceanic lithosphere and began to plunge beneath southern Eurasia until its buoyancy resisted subduction.

Just as oceanic lithosphere flexes down at island arc structures to form a deep-sea trench (Figures 5, 21, and 22), the Indian lithosphere bends down in front of the Himalaya to form a deep basin. Sediment eroded from the Himalaya has filled that basin, the Ganga Basin (Figure 31). The thickest accumulation lies at the foot of the mountains, and thins southward where it laps onto the old rock of India. The modern Ganga or Ganges River flows along the basin, with tributaries from the Himalaya to its north depositing sediment as they flow toward the Ganga.

The Ganga Basin, roughly 300 km in width, is notably wider than the trenches along the margins of the Pacific Ocean (which are more typically 100–150 km in width: Figure 22), apparently because the Indian lithosphere is thicker than typical oceanic lithosphere. Just as a thick sheet of paper hanging over the edge of a table bends less than a thin one, so does thick lithosphere. Because thick paper and thick lithosphere are less flexible than thin versions, they can support heavier loads where they are bent down—more paper clips hooked to the former and higher mountains atop the latter. The Indian lithosphere is also thicker than that under Europe to the north of the Alps. It follows that one reason that the Himalaya hosts the highest mountains on Earth, much higher than those in the Alps, might be that the thick, stiff Indian lithosphere bends less than the lithosphere thrust beneath other mountain ranges, such as the thinner, more flexible European lithosphere that has been thrust beneath the Alps.

The Tibetan Plateau

Farther north, the Tibetan Plateau has undergone a history quite different from that of the Himalaya. Most think that before India collided with Eurasia, and before the last ocean floor between them vanished, southern Asia was bounded by a continental margin similar to that of the present-day Central Andes. Volcanoes erupted onto high terrain along the southern edge of Eurasia, as they do in Peru, western Bolivia, and Chile today, and granite intruded at depth beneath the volcanoes. As noted above, after the northern margin of India met the southern edge of Eurasia, and northern India began to plunge beneath southern Eurasia, the buoyancy of India's crust resisted continued subduction. India's lower crust and mantle lithosphere continued to slide beneath the southern edge of Eurasia, but half of India's convergence has occurred by its penetrating into Eurasia. As India moves northward, it steadily pushes, or drags, southern Eurasia's ancient Andes-style mountain range northward into Eurasia. The crust of this region shortened and thickened as it was compressed, and a high wide plateau, the Tibetan Plateau, grew. The buoyancy of thick crust under Tibet, in isostatic equilibrium, supports the high Tibetan Plateau, just as thick crust buoys up the Central Andes.

Earthquakes occur throughout the Tibetan Plateau, and active faults slice its upper crust into small fragments. If one were to describe how each fragment of crust moved with respect to others, as we do for plates in plate tectonics, one would need to consider hundreds, if not thousands, of fragments. Such a description is so unwieldy as to be useless. Although some scientists do try to describe crustal movements in Tibet in terms of separate fragments, most, like me, prefer to imagine that the Tibetan lithosphere, including both its brittle crust and a thin mantle part, deforms as a viscous fluid. The thick crust with low strength, over a weak uppermost mantle, creates an image like honey or molasses overlain by crumbs, fragments of stiffer upper crust. Gravity acting on the honey or molasses that has spilled onto a dish or table-top causes it to spread outward and carry its upper-crustal crumbs with it. The same occurs within Tibet today; not just crust, but presumably the entire lithosphere, beneath the highest parts of Tibet currently spreads apart and becomes thinner, as its southern margin spreads onto the India plate.

An analogy is sometimes made with ripe Camembert or Brie cheese. When ripe, such cheese spreads out over the plate on which it was placed (or the India tectonic plate in Tibet's case), and it (as well as Tibet's crust) thins. Unfortunately, the analogy with honey, molasses, or ripening cheese is imperfect, because most of Tibet overlies weak asthenosphere, not a rigid plate. Thus, resistance to flow of the honey-like or cheese-like continental lithosphere is less than that of honey, molasses, or ripening cheese flowing over a table-top or plate.

The Tien Shan and Mongolia

The effects of India's penetration into Eurasia do not stop at the northern edge of the Tibetan Plateau. An effectively rigid block (or plate) underlies the Tarim Basin just north of Tibet (Figure 34). The Tarim Basin is a large, relatively low area that contains the Taklimakan, a huge desert that is virtually uninhabited except along its margins. Behaving like a small plate, Tarim moves northward relative to Eurasia, by rotating about an axis near its eastern end. On its northern side, the Tien Shan mountain belt forms a wall that extends from east to west and contains the only peaks higher than 7000 m outside the Himalaya and its continuations east and west (Figure 34). The high mountains on the margins of the Tarim Basin collect water that trickles down to the oases along the northern and southern branches of the Silk Road.

The east–west Tien Shan has grown by the north–south shortening and thickening of crust between the Tarim Basin and the stable, strong Kazakh Platform of the Eurasia plate. The construction of the Tien Shan, however, differs from that of the Himalaya, where one plate, the India plate, followed oceanic lithosphere into a subduction zone and continues to slide under the high mountain belt. In the Tien Shan, north–south shortening and thickening of the crust occurs across the entire width of the belt. This is accomplished by slip on many thrust faults, not just one beneath each margin of the mountain belt, but also on others within the belt. Because such faults dip either north or south, in some places basins have formed where rock on their northern and southern sides has been thrust up to build mountain ranges. The weight of that rock in the mountain ranges that has been thrust over the basins pushes the floors of the adjacent basins down by amounts comparable to the heights of the ranges. If sediment cannot accumulate in a basin fast enough, its surface can remain low, and even drop below sea level. The world's second lowest place on land, the Turfan Depression near the eastern end of the Tien Shan, formed in response to the weight of mountain ranges to its north and south. (The lowest place, containing the Dead Sea between Israel and Jordan, formed by the north and south sides pulling apart obliquely, and sediment has accumulated too slowly to keep pace with the subsiding basin floor).

Northeast and east of the Tien Shan, northeast–southwest crustal shortening occurs across the Mongolian Altay in western Mongolia and the Gobi-Altay in southern Mongolia (Figure 34). We see all of this ongoing deformation as the result of India's penetration into Eurasia. As in the Tien Shan, many faults are active, but unlike the Tien Shan, strike-slip faulting is widespread—right-lateral on northwest-trending planes in the Mongolian Altay and left-lateral on east-trending planes in the Gobi-Altay. Crudely, the north-northeastward penetration of India into Eurasia is wedging Mongolia apart. Additional evidence for such a wedging apart comes from the area northeast of this region, where Lake Baikal, the deepest lake in the world, occupies a major rift zone (Figure 34). In the Baikal Rift, northwest–southeast divergence of crust has created the deep rift valley.

This southeastward movement of material away from the rest of Eurasia is not restricted to the region surrounding the Baikal Rift, but applies to all of eastern Asia. Specifically, the eastern parts of Siberia, Mongolia, and China all move southeast to east-southeast with respect to the rigid Eurasia plate (Figure 36). For example, South China moves almost 10 mm per year east-southeast away from rigid Eurasia. This eastern region does not behave as a single rigid plate, however, for South China and North China are separated by a wide zone of deformation, where many destructive earthquakes have occurred, including one in 1556 that took 830,000 lives. Velocities of GPS points show that the eastern part of

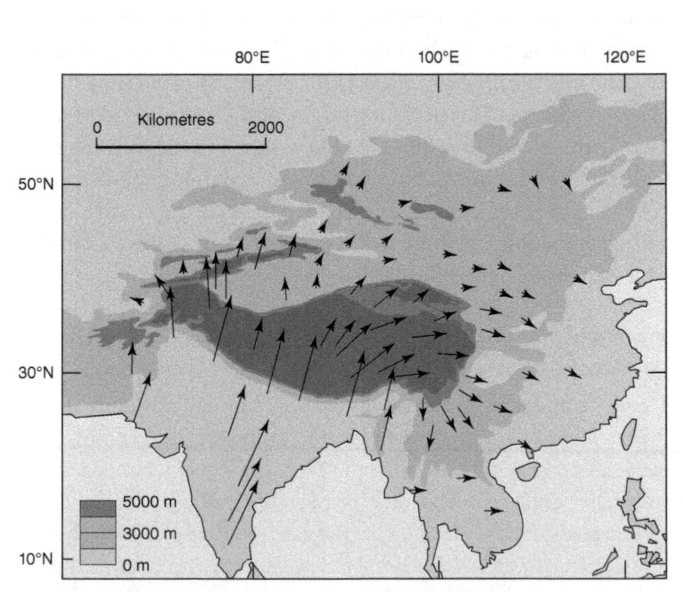

36. Map of eastern Asia with arrows showing velocities of selected GPS control points with respect to a fixed Eurasia, the northern part of the map.

the Tibetan Plateau flows around the northeast corner of the Indian subcontinent as it penetrates into the Eurasian landmass (Figure 36).

In 1922, the Swiss geologist Emile Argand proposed that a wide northern part of India had slid beneath Tibet, and this movement had caused deformation over a broad area of Asia. This suggestion followed on the heels of Wegener's proposal that continents had drifted apart and, as with continental drift, few seemed to have paid much attention to Argand's inferences. In 1975, following the wave of enthusiasm for plate tectonics, the French geologist Paul Tapponnier and I again suggested that all of this widespread deformation in Asia resulted from India's penetration into Eurasia, and our work was greeted with enthusiasm.

Other regions of continental deformation and mountain building

The various phenomena I have described are not unique to Asia. As noted, the style of deformation described for the Himalaya (Figure 31) applies also to the Alps. Layers of sedimentary rock deposited on southern Europe's ancient continental margin, thick layers of limestone, sandstone, and shale, were scraped off Europe's margin as the Eurasia plate plunged into a subduction zone, and these layers were folded and stacked atop one another. With further convergence, some of Eurasia's stronger, deeper metamorphic and igneous rock was also sliced off, as the remaining underlying lithosphere penetrated deeper into the subduction zone under northern Italy. That convergence, which probably was never as rapid as in the Himalaya, seems to have slowed dramatically, if not stopped, 5–10 million years ago.

Tibet is not unique among high terrains undergoing collapse and outward spreading. The same has been happening for tens of millions of years in the 'Basin and Range Province' of the western United States, in Nevada, western Utah, and surrounding regions (Figure 37). The east–west dimension of the region has approximately doubled in width. Some 20–40 million years ago, the region between Sierra Nevada in California and the region now occupied by the Great Salt Lake seems to have been a high plateau, resembling the Central Andes, perhaps 4000 m high and approximately 500 km wide. After a long period of spreading apart and collapsing as the underlying crust thinned, Nevada and Utah now present a much lower (~1500 m) but wider (~1000 km) region. Because the upper crust does not flow, as the lower crust and upper mantle do, it has broken into blocks (like crumbs) of crust, which make for north–south-trending ranges and adjacent basins that give the region its name, the Basin and Range Province.

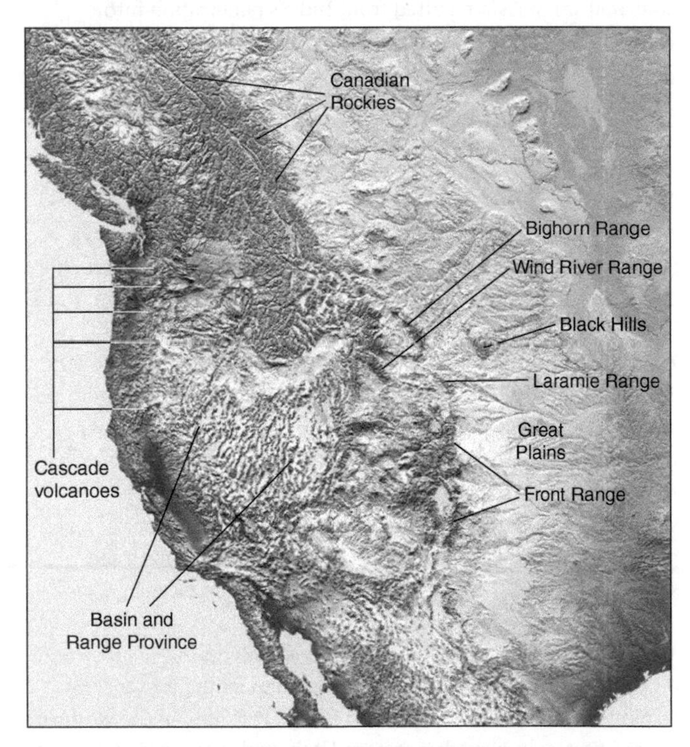

37. Shaded relief map of the western USA showing regions with different styles of deformation.

When plate tectonics was first recognized, the Rocky Mountains of western North America immediately posed a puzzle. It was soon recognized that the high terrain of western Canada developed along a subduction zone, like that along the present-day Andes. Seafloor west of North America plunged eastward beneath the west coast. Subduction stopped some 30 million years ago, but erosion has not yet destroyed the high terrain. In particular, the processes that elevated the beautiful mountainous terrain in Canada's Banff and Jasper National Parks operate today on the east side of the Central Andes, in southern Peru, Bolivia, and northern Chile. The flat stable terrain of most of the Canadian interior was thrust westward beneath the Canadian Rockies, just as the interior of South America in Brazil and Argentina today is thrust westward beneath the Central Andes. Layers of sedimentary rock deposited on the flat terrain became folded and thrust atop one another (Figure 33), in the same manner that similar sedimentary rock is currently being folded as the South America plate is thrust westward beneath the Andes.

The Southern Rocky Mountains of the western United States have undergone a different history from the Rockies of Canada. First, the southward continuation of the Andes-like range in Canada developed west of the Southern Rocky Mountains (Figure 37). As discussed above, that Andes-like belt, including what is now the Basin and Range Province, apparently once was a high plateau that later spread apart and collapsed. The Rocky Mountains of the western United States, however, lie deep within the North American continent, east of the present-day Basin and Range Province, far inland from the subduction zone that lay along the west coast of North America from 150 to as recently as 30 million years ago. Unlike the Himalaya, there is no single thrust fault on which the Southern Rocky Mountains have been thrust onto adjacent strong lithosphere. Instead, faulting similar to that currently happening within the Tien Shan seems to have built the many separate ranges, the Bighorn, Wind River, Laramie, and Front Ranges and even the Black Hills (Figure 37) that collectively comprise the Rockies. Although no continental collision like that in the Himalaya occurred along the west coast of North America, the compression of the western edge of North America not only built a high plateau, like those in the Central Andes and Tibet, but also induced compression of terrain farther east, deep within North America in Colorado, just as the pressure India applies to southern Eurasia induces compression in the Tien Shan deep within Eurasia.

The analogy of the Rocky Mountains with the Tien Shan is imperfect, however, because the thrust faulting some 70 to 50 million years ago cannot explain the present-day elevation of the Southern Rockies. Such crustal shortening and thickening can account for only part of the high (~2000–3000 m) mean elevations of the Rockies, and it cannot explain any of the 1500 m elevation of the Great Plains east of the Rockies. Denver lies at 1600 m today, but has been built on rock that was below sea level 70 million years ago. The whole region is in isostatic balance, buoyed up by light material beneath it. Unlike the Tibetan Plateau or the Central Andes, where thick crust provides compensating deficit of mass, however, the compensating deficit of mass at depth beneath the Rockies seems to lie largely within the uppermost mantle. Either the uppermost mantle is anomalously hot, as it is beneath mid-ocean ridges, and/or it has become less dense because some dense minerals have metamorphosed into less dense minerals. In either case, the process that has made the Rockies and the Great Plains to its east stand high is not a part of plate tectonics, but a complication that makes plate tectonics incomplete.

In summary, the tectonic processes that have shaped continents differ from plate tectonics, which applies largely to oceanic regions. In particular, the much greater thickness of continental than oceanic crust (commonly 35 km or more for the former and only 7 km for the latter) makes continental and oceanic lithosphere behave differently.

First, because crust is less dense and therefore buoyant, compared with the mantle, thick continental crust resists subduction into the asthenosphere. Where a continent, as a passenger on a larger plate of lithosphere, follows oceanic lithosphere into a subduction zone, the thick continental crust causes subduction to choke. Although the analogy of a subduction zone with an oesophagus is inadequate, in both cases swallowing is made easier by breaking the material being consumed into smaller pieces. The consequence for subduction of continental lithosphere commonly is for slices of the upper part of the crust to detach from underlying parts. These slices then become stacked atop one another to form a mountain range, like the Alps or Himalaya. The deeper lithosphere stripped of some of its buoyant crust seems to continue to plunge into the asthenosphere.

Second, except possibly for the oldest, coldest lithosphere, continental lithosphere seems to be weaker than oceanic lithosphere. For this reason, plate tectonics works best in oceanic regions. When put under stress, oceanic lithosphere can remain strong enough to behave like an effectively rigid plate, but continental lithosphere will deform. Whereas plates of oceanic lithosphere are separated by narrow belts of earthquakes, which are themselves manifestations of relative movement between adjacent plates, earthquakes occur across vast portions of continents. Some would argue that continents are simply broken into many microplates, or blocks, but in such regions there are so many blocks that keeping track of all of them becomes too unwieldy to be helpful. For many aspects of intracontinental deformation, but not all, a more useful approach is to treat continental lithosphere as a viscous fluid, like honey or molasses, if capped by crumbs of brittle upper crust.

Many mountain ranges seem to have been built by widespread deformation of continental lithosphere. When the horizontal dimension of a region of continental crust is shortened, the crust thickens. Because of isostasy, thick buoyant crust stands higher than thin crust. No analogous process occurs in oceanic lithosphere, and this process is not part of the plate tectonics canon.

Chapter 7
From whence to whither?

By the early 1970s, the basic ideas of plate tectonics—rigid plates created at mid-ocean ridges, sliding past one another at transform faults, cooling and subsiding as they age, and eventually plunging back into the asthenosphere at subduction zones—had passed enough tests that scientific questions associated with the basic idea no longer lay at the forefront of Earth Science. Plate tectonics, of course, did not die, or become moribund, but rather became the foundation for other subjects whose significance in many cases had not yet been appreciated. In this last chapter, I discuss a couple of examples that illustrate the role of plate tectonics in questions that related to society and science, including one that seems to remain open, and I end with a personal view of how plate tectonics affected the way we approach questions in Earth Science.

Recurrence of great earthquakes

In 1835, Charles Darwin was carrying out geologic fieldwork in Chile when a huge earthquake occurred. In its aftermath he found islands that had risen several metres, and farther inland he found regions that had subsided metres, in a pattern similar to that mapped by George Plafker after the 1964 Alaskan earthquake (described in Chapter 4). In 2010, 175 years later, another earthquake of comparable magnitude occurred in essentially the same part of Chile. The 175-year interval is noteworthy.

The seafloor west of Chile, part of the Nazca plate (Figure 5), moves east toward South America and plunges beneath the coast of Chile at a rate of approximately 70 mm per year, or 7 m per century. During great earthquakes like those in 1835 and 2010, slip of approximately 10 m occurs; the western edge of the South America plate lurches 10 m over the edge of the Nazca plate. Imagine putting a block of wood on the floor, attaching a compressible spring to its side, and then pushing on the spring. At first the spring contracts, but the block does not move. Then when the spring is compressed enough, frictional resistance is overcome, the block lurches across the floor, and the spring becomes less compressed. The spring and block are analogous to the coast of Chile, and the floor to the Nazca plate. The steady motion of the Nazca plate toward the South America plate loads the Chilean coastal spring until friction on the boundary between them can no longer resist slip on that boundary. With the steady convergence of the Nazca plate with the South America plate, this slow, steady compressing of the Chilean coastal spring followed by an abrupt lurching of the west coast of Chile over the Nazca plate will occur time and again.

In the late 1990s, a group of Chilean and French geologists reasoned that in the 160 years since the earthquake that Darwin had studied, 11 m of potential slip had accumulated (160 yr × 7 m/century = 11.2 m), and a repeat of the 1835 earthquake might be imminent. They added new GPS control points to their network of GPS stations, and intensified study of the region in anticipation of a major earthquake. Although there was no way to predict the decade, let alone the year or the day, that the 2010 earthquake would occur, the knowledge that such an earthquake in that part of Chile was likely can be seen as a social benefit of plate tectonics.

Similar logic underlies expectations of the San Andreas fault in California. Approximately 4 m of slip (Figure 17) occurred in the earthquake in 1906 that damaged San Francisco so badly. Resurveying of control points using both classical, 19th century techniques and GPS measurements suggests that the two sides of the fault slide past one another at approximately 25 mm/yr, or 2.5 m per century. The ratio of 4 m to 2.5 m per century gives an approximate recurrence interval of 160 years, making the decades surrounding 2066 seem ominous.

Plate tectonics and glaciation

Plate tectonics affects regional climates by shifting landmasses, which in turn affect circulation of the atmosphere and ocean. For example, as Wegener recognized, some 300 million years ago all of the major continents of the southern hemisphere—Antarctica, Africa, South America, Australia, and also India—not only were grouped together as one big continent, Gondwanaland, but for a period Gondwanaland was centred on the South Pole, as Antarctica is today. Geologic evidence of past glacial activity, in the form of sediment typical of that carried by glaciers, can be found on virtually all of today's modern continents.

Plate tectonics not only shifts large landmasses long distances, but it also closes and opens gateways between continents, with the result that ocean circulation can be blocked, or unblocked. Unlike the atmosphere, which can flow around mountains, water in the oceans cannot cross topographic barriers above sea level. Today, water in the Atlantic Ocean cannot get to the Pacific except through the Southern Ocean. (Water flows from the Pacific to the Atlantic through the Bering Straits between Asia and North America, not in the opposite direction.) Warm water from the central Atlantic, and in particular from the Caribbean Sea and Gulf of Mexico, flows northward carrying heat with it. When that water cools, it sinks deep into the North Atlantic and then most of it exits the Atlantic through the depths of the South Atlantic. That deep water is thought to reach the surface again after approximately 1000 years, and then mostly in the North Pacific.

Several million years ago, however, North and South America were separate; in place of the Isthmus of Panama was the Central American Seaway though which water from the Pacific is thought to have passed into the Atlantic. Accordingly, ocean circulation in the Atlantic could have been very different from that today. Many, though not all and not I, associate the emergence of the Isthmus of Panama with the onset of recurring Ice Ages, when ice sheets spread over Canada and over much of Finland and Scandinavia (Fennoscandia).

Prevailing winds over the ocean drive a circulation that includes strong poleward currents along the western margins of oceans. The Gulf Stream is a familiar example, and the Kuroshio Current along the coast of Japan is another, but the strongest is the Agulhas Current along the east coast of Africa. These currents contribute to the poleward transport of heat to high latitudes. Antarctica, however, is surrounded by ocean, the Southern Ocean, which is famous for its strong winds and rough seas. A strong eastward-flowing current, the Antarctic Circumpolar Current, carries water rapidly around Antarctica, so that water from more tropical regions becomes drawn into this current before it can reach Antarctica itself. In a sense, by isolating Antarctica from this warm, poleward flowing water, the Antarctic Circumpolar Current insulates Antarctica and keeps it cold.

Forty million years ago, Australia lay near the coast of Antarctica, and the island of Tasmania and the shallow seafloor to its south lay against the east coast of East Antarctica (Figure 38), near McMurdo Bay where bases for Antarctic exploration were established more than 100 years ago. Water in the southern Indian Ocean could not flow south of Australia and into the South Pacific. At the same time, the southern tip of South America and the Antarctic Peninsula also lay against each other, and water could not flow from the Pacific to the Atlantic. The Antarctic Circumpolar Current was blocked in two places. Approximately 30–35 million years ago, these gateways opened. Shortly

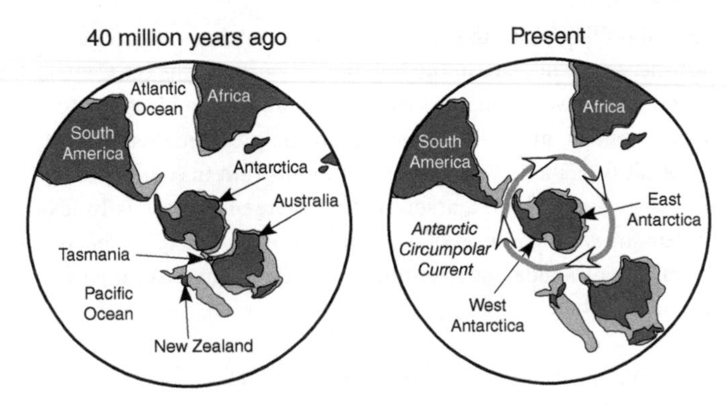

38. Maps of the Southern Ocean and surrounding continents 40 million years ago (left) and today (right).

afterwards, ice abruptly covered East Antarctica. Many believe that the opening of these gateways and the development of the Antarctic Circumpolar Current insulated Antarctica and enabled the ice sheet to develop there.

Petroleum resources and ore deposits

Geology's greatest impact on society surely derives from resources that the Earth provides, such as oil and gas or iron, aluminum, and copper, as well as rarer metals used in sundry ways. We might ask: Has plate tectonics facilitated the discovery and acquisition of such resources?

Consider oil and gas first. Several conditions must be met for oil or natural gas to accumulate in the Earth. First, organic material—trees, grass, bushes, as well as microorganisms—must be buried. Burial must occur by sufficiently rapid accumulation of sediment on top of the organic material before its decomposed products enter the atmosphere. Part of the capping sediment must be fine-grained impermeable sediment, so that the oil and natural gas cannot leak to the surface. Finally, heat is needed to accelerate decomposition, but of course too much heat can destroy the organic material either before or after it becomes oil or gas.

Newly formed continental margins provide ideal, but not unique, environments for petroleum resources. When a large continent splits into two separate smaller continents, and a new ocean grows wider between them, first a rift valley, like that in East Africa, forms. During this rifting phase, not only the crust, but also the underlying mantle lithosphere become thinner. Some sediment accumulates in the rift valley, but more important is the subsequent slow development of the continental margin after the rift valley splits and new oceanic lithosphere forms in the space between the diverging continents.

Because of the rifting, relatively thin crust underlies the new continental margins. Because of isostasy, the top of this thin crust lies below sea level to form a continental shelf, and seaward of the continental shelf lies deeper seafloor that forms by seafloor spreading. In tropical regions, coral reefs, made largely of calcium carbonate, can develop, and other organisms can also thrive. Sediment brought from the adjacent land mass not only accumulates on both the shelf and the deeper adjacent seafloor, but it can bury organic material. Where relief near the margin is modest, rivers flow with low gradients, and such rivers can carry only fine-grain sediment, like mud. Mud, when it consolidates into sedimentary rock, becomes shale, which is especially impermeable. Finally, because of the thin lithosphere at the new margin, the uppermost mantle beneath the thin crust is unusually warm. Thus, more heat is conducted through the crust than through older, colder continental lithosphere. That heat can accelerate the maturation of organic material into oil and gas.

A particularly ideal time and place for petroleum resources occurred in the Mesozoic Era, when the Earth was warm and the concentration of carbon dioxide in the Earth's atmosphere was apparently high. So, life flourished, and much organic material was deposited. At that time, also, apparently the northeast margin of the Arabian subcontinent was formed by the rifting away of another continental fragment whose present-day whereabouts is not known. This occurred near the equator, where water was especially warm. This setting—a new continental margin at a time with a warm climate and abundant carbon dioxide—enabled organic material to accumulate in abundance, and then to become oil along the Persian Gulf. The Arabian subcontinent later, approximately 35 million years ago, collided with southern Eurasia to form the Zagros Mountains of southwestern Iran. The rich petroleum reserves in that region lie beneath both undeformed sedimentary rock along the Persian Gulf and its southwestern margin, and beneath folded limestone and sandstone that comprise the Zagros Mountains.

Petroleum was discovered long before plate tectonics, and obviously an understanding of plate tectonics played no role in the discovery of these resources. Nevertheless, geologists in modern oil companies routinely employ the basic concepts of plate tectonics to make decisions about where to explore for oil. In 1974, an eminent petroleum geologist, Peter Vail from Exxon, was visiting. Having studied during the plate tectonics era, I told him that I must know something that could contribute to the discovery of oil. So, I asked him what I should do. He replied, 'Just keep doing what you are doing.'

Valuable metals, like copper, gold, and silver, pervade the crust, the mantle, and the ocean too, as isolated atoms or as constituents of minerals, but only in minuscule concentrations, large enough to be measured with sensitive equipment, but not likely to be commercially viable. Only in those few places where they have become concentrated does the cost of mining them merit effort. Although different metals, and different minerals, require different conditions for their concentration, a common theme includes both water, which can flow easily through cracks in rock, and relatively high temperatures that accelerate dissolution of elements and chemical compounds from the rock. Thus, in ideal settings, metals either in pure forms or in ore-grade minerals can become concentrated by fluids percolating through the crust. Many such ore deposits are associated with magmas, molten volcanic rock that intrudes the crust but does not reach the surface.

In the context of plate tectonics, most regions where volcanism occurs, and therefore where magma is present below the surface, lie along mid-ocean ridges or beneath volcanoes at subduction zones. As mid-ocean ridges are not easily reached, particularly by large apparatus, they do not yet offer profitable targets for commercial mining. In many, but far from all, cases, mining of copper, gold, and silver as well as of lead, tin, molybdenum, and tungsten is carried out in settings like the Andes and similar regions where subduction zones beneath a continent had occurred earlier in geologic history.

As discussed in Chapter 4, the subduction of oceanic crust carries water, mostly bound to minerals and not in its liquid form, into the mantle. The presence of hydrogen in crystal lattices lowers the melting temperature of mantle minerals, and magmas form at a depth of approximately 100 km. In addition, some sediment that is carried down at the subduction zone also melts at a relatively low temperature, and this molten material helps to gather magma into quantities large enough that it can rise through the overlying mantle and into the crust. Moreover, at least some valuable metals preferentially make their way into the magma rather than lingering in the remaining solid crystals. Thus, the rising magmas become enriched in such metals. As the magmas rise through the crust, they may also extract valuable metals from the crust, but many such metals seem to be derived largely from the mantle. Then, when magmas coalesce beneath volcanoes, they offer a source of metal, though not yet of sufficiently high grade as to merit exploitation, and a source of heat to drive fluid flow through the overlying crust. Finally, for such ore deposits to be exposed, they must be exhumed, and erosion is far more effective in mountainous territory than where islands barely poke above sea level.

Thus, it should be no surprise that the Andes host many of the world's major ore deposits. Similarly, the gold that triggered California's 'Gold Rush' has come from terrain that 100 million years ago occupied a setting similar to that of the present-day Andes.

In the late 1970s, an outspoken colleague of mine, John Edmond, was visiting a nearby prestigious university where an annual celebration was being held in honour of an eminent ore geologist. The geology faculty in that institution had been slow to accept plate tectonics, and Edmond was using that fact to taunt some of the graduates, those much older than he and quite wealthy from having found economically viable ore deposits. Edmond pushed too far, and finally one of them said, 'And how many ore deposits have you found with your plate tectonics?' Most would argue today that the understanding that plate tectonics brought of how the Earth behaves has been vital to both the oil and the ore industries, but I suspect that no major discovery could be attributed to the essentials of plate tectonics.

Plate tectonics on other planets

Plate tectonics was recognized at the same time that exploration of the solar system began, and planetary scientists of course were curious to learn whether other planets exhibited plate-like behaviour. It was known that the outer planets, Jupiter, Saturn, Uranus, and Neptune, are not solid, but consist of gases. So, the question focused on the inner, solid planets, Mars, Venus, Mercury, and the Moon.

Recall that the key to plate tectonics is the lithosphere, the cold strong outer layer through which heat is conducted from the

Earth's interior to its surface. For subduction of lithosphere to occur, a plate must bend down to form a deep-sea trench, before it can slide beneath another plate. It is easy to imagine that if the lithosphere were too thick, it might not be able to bend sufficiently, so that subduction would be impeded. This seems to have happened on the Moon, Mercury, and Mars. We often describe these as one-plate planets.

All three are much smaller than the Earth. We presume that the rate that heat will be produced in a planet, by the decay of radioactive elements, will scale as the volume of the planet, which varies with the cube of its radius. The rate that heat is lost, however, depends on the surface area of a planet, which scales as the square of its radius. Thus, large planets will gain or keep heat longer than small ones, and conversely small planets will cool faster than large warm ones. Lithospheric thicknesses should be greater on the small planets.

Venus is nearly the same size as the Earth. So, we might expect these two planets to behave similarly. Venus, however, differs from the Earth in an important way; its atmosphere is thick and opaque, with abundant greenhouse gases. The temperature at the surface is more than 450°C! So, its hot crust should be weak, and the mantle part of its lithosphere thin. Perhaps, therefore, it is no surprise that the surface of Venus shows little evidence of rigid plates, but plenty of topography resulting from deformation of its crust. One might say that Venus exhibits continental tectonics on steroids.

The birth of plate tectonics

Most agree that the Earth formed by the collisions of small planetesimals, small solid objects that condensed from a nebular cloud before accreting together to form the Earth, some 4.5 billion years ago. As the planetesimals accreted to the growing Earth, the energy that they lost on impact was converted to heat. Soon the outer part of the Earth, if not all of it, melted, and was covered by a sea of lava. Plate tectonics had to wait for the lava-covered Earth to cool off.

At the opposite extreme, we can be confident that plate tectonics had begun by 200 million years ago, the age of the oldest ocean floor, and therefore the oldest oceanic lithosphere that is present today. Obviously, 4.5 billion and 200 million years do not place tight bounds on when plate tectonics began. As is often the case with questions lacking convincing answers, opinions not only differ but can also be strong.

Most geologists who consider this question turn to 'geological corollaries', geologic observations that seem to be associated with plate tectonics and that can be seen in the older geologic record. Two particular rock types seem intrinsically linked to plate tectonics: belts of granite and 'ophiolites'.

Granite underlies the volcanoes at island arcs, and hence at subduction zones. Not all granite formed at subduction zones, but granite bodies that did commonly bear chemical signatures that allow them to be identified. Among prominent examples of such granitic belts, the Sierra Nevada in California (Figure 37) serves as one good example. Oceanic lithosphere plunged eastward beneath the Sierra Nevada from before 150 million years ago to as recently as 30 million years ago (see Chapter 5). Another prominent granitic belt follows the southern edge of Tibet, just north of rock that was part of India until it collided with Eurasia approximately 50 million years ago, as discussed in Chapter 6.

Ophiolites, first found in the Alps and recognized as ancient ocean floor, form belts parallel to granitic belts in many regions. A complete ophiolite suite includes: (1) marine sediment consisting of fossil plankton called radiolarian that had been deposited in the deep ocean, (2) basalt that shows textures implying that it had cooled rapidly under water, and (3) serpentinite, a green metamorphic rock that formed by the addition of water to peridotite, the rock constituting typical upper mantle. The word ophiolite derives from the Greek words *ophis* meaning 'snake' and *lithos* meaning 'rock'. The scaly texture of serpentinite reminded geologists of a snake's skin. Ophiolites commonly crop out on the Earth's surface as jumbled mixture with little coherence. Their significance, beyond having an origin in the deep ocean, remained poorly appreciated until with plate tectonics it became clear that ophiolites could be used to study oceanic crust in ways that marine geologists working from ships at the sea surface could not.

Ophiolites commonly crop out where two continents have collided and been sutured together. One of the classic examples is a narrow belt that marks the boundary between sedimentary rock that defined the northern edge of India before India collided with southern Tibet, and the rock, mostly granite, that lay along the southern edge of Eurasia. In 1936, the Swiss geologist Augusto Gansser snuck across the Tibetan border with India, disguised as a Buddhist pilgrim en route to Mount Kailas, sacred to both Hindus and Buddhists. While en route, he mapped the ophiolites nearby. As Gansser recognized, the belts of ophiolites and granite that follow the northern side of the Himalaya mark the suturing of India to Eurasia. As a second example, rock cropping out in the Coast Ranges of California includes ophiolites, which were presumably scraped off the seafloor that plunged beneath the Sierra Nevada when granite formed in that region.

If ophiolites and granitic belts defined ancient subduction zones and marked sutures between collided continents, determining when plate tectonics began could be reduced to finding the oldest such belts. Indeed, evidence of both ophiolites and subduction-related granite belts has been reported from rock as old as 3.8 billion years, which, if credible, would suggest that plate tectonics began a few hundred million years after the Earth formed. A few qualified sceptics, however, question both the inferences of ancient ophiolites and the association of the ancient granites with subduction. Some claim that clear ophiolites cannot be found until 2.5 billion years ago, and perhaps not until 1.5 billion years ago.

The question is complicated by ophiolites being an aberration, an imperfection in plate tectonics. If subduction occurred smoothly and simply, granite belts would form, but all oceanic crust would be subducted. We would not have ophiolites to mark sutures, because all of their ingredients would have been carried deep into the mantle at subduction zones. 'Cryptic sutures' would be the norm. Indeed, ophiolite belts are rarely continuous along suture zones, and it seems possible that ancient sutures have yet to be found.

The differences of opinion were made clear at a meeting in 1975, organized by John Dewey, who more than most geologists had used plate tectonics to interpret the history of continents, and is now retired from Oxford University. The meeting was to address evidence for plate tectonics before approximately 200 million years ago. Two leaders of geology, Robert Shackleton and Kevin Burke, argued the question throughout the meeting. Shackleton, grand-nephew of the explorer Ernest Shackleton and already in his mid-60s, had walked over vast stretches of rock throughout his life; if some tectonic process were revealed in the rock record, he would have seen it. Burke, twenty years Shackleton's junior, brought an ability to synthesize geologic observations, made either by others or by him, in some cases using images of the Earth from satellites. Their differences were summarized by someone else who said, 'Shackleton could not see evidence that an ocean basin had closed and two continents had collided until he heard the sound of the waves, while all that Burke needed was one fuchsite crystal.' (Fuchsite is a rare mineral found in ophiolites.) When this remark was made, someone from the back of the room yelled, 'Yeah, and seen from a satellite.'

Much has been learned in forty years, but when rigid plates formed and plate tectonics dominated the evolution of the Earth remains a hotly argued question.

Earth Science in general

My father taught me a basic tautology: when something seems complicated, we do not understand it, but when we do understand something, it has become simple. Plate tectonics brought a simple, easily understood concept—plates of lithosphere, made at spreading centres, moving over the surface of the Earth as rigid objects, and eliminated at subduction zones. Whereas to a large extent inferred descriptions of complicated interactions of many processes had sufficed before the recognition of plate tectonics, what followed were increasingly frequent attempts to isolate processes that could be understood and then to build a foundation on which to increase and broaden comprehension.

As the implications of plate tectonics diffused through Earth Science, in most cases they fostered an appreciation of simplicity where complexity previously had reigned. Large-scale processes, those affecting huge regions like continents, or even mere mountain ranges, had been the domain of speculative thinkers. Few geologists had been trained to look at large regions. Yet, we now know that if we could study a new planet, we would start with the large scale, and work down to the smaller scale, rather than the opposite. Of course, the birth of satellite-based remote sensing, which occurred concurrently with the recognition of plate tectonics, facilitated this appreciation for the large scale. Not only did it make sense to examine whole regions at once, but also we had the new technology, not merely seismology or sparse ships' tracks across the open ocean, to do so. Obviously, the study of the Moon and planets has relied almost entirely on remote sensing to reveal both the shapes of their surfaces and their deeper structures. Moreover, the focus on large scales did not supplant the focus on smaller scales, but instead it brought new problems that required an integrated understanding of how processes occurred at all scales.

The recognition of plate tectonics united geologists of all stripes, from those doing most of their work in the field with their boots on to those in armchairs working with remotely sensed observations, and knowing far less about dinosaurs than do most 6-year-olds. The result was a focus on processes. Of course, many geologists before plate tectonics concerned themselves with processes. One of the most fundamental dictums taught to geology students is that 'The present is the key to the past', and one of its leading early proponents, Charles Lyell, wrote in the 1830s that understanding the processes that have shaped the Earth should be one of geologists' main goals. With the recognition and testing of plate tectonics, which has relied on quantitative aspects (rates, directions, and amounts of relative movement, depths of oceans, rates of heat loss from the Earth, etc.), the focus shifted more from the description of phenomena that had occurred as part of a history of the Earth, to an understanding of the underlying processes. Although geologists still construct, if not concoct, stories to describe the geologic histories of separate patches of ground, typically those stories not only are aimed at testing hypothesized processes, but also require quantitative measurements to test their essential elements.

Most obviously, plate tectonics included continental drift, and those who, for whatever reason, thought that large horizontal displacements of continents were impossible were forced to change their views (or dig their heels into arguments with little foundation). To some extent this is ironic, for plate tectonics describes so well the evolution of oceanic lithosphere, but fails in many ways for continents, the object of study for most geologists. Moreover, the confirmation of continental drift with geophysical techniques, whether from ships at sea or seismographs recording distant earthquakes, raises the question of how so many, but by no means all, geologists failed to see in their outcropping rock the evidence for such a grand unifying idea. The answer is simply that the rock record on land offers a poor record of a large-scale process like plate tectonics. As James Jackson of Cambridge University likes to say, the area affected by an earthquake with a Richter magnitude of 6 affects a region larger than that which a graduate student can map for a PhD thesis, and each year hundreds more such earthquakes of that size occur than PhDs in geology are granted.

Most scientists do not accept a new idea until they see how it affects the data that they themselves gather and analyse. Accordingly, as the understanding of plate tectonics made its way through Earth Science, different communities accepted the idea at different rates. For example, palaeontologists, who study how plants and animals have emerged, evolved, and become extinct over millions of years, were quick to accept, and use, plate tectonics. For them continental drift had offered a sensible, but not unique, explanation for why some plants or animals started to evolve differently on diverging continents.

For petrologists, who study how different kinds of rock form, plate tectonics brought obvious corollaries. It was only in the 1960s that it became obvious that the ocean floor was underlain by basalt, the rock that forms when mantle material melts at pressures corresponding to depths of 10–30 km. The temperatures at which most rock-forming minerals melt decrease with decreasing pressure, and therefore with decreasing depth. As the two plates diverge at mid-ocean ridges, hot mantle material rises passively to shallow depths without cooling much. Thus, the temperature of that rising rock, without changing much at all, passes from being too cold to melt to too hot not to melt. Although all of the logic given above was known before plate tectonics was suggested, the making of oceanic crust became a simple corollary of plate tectonics. Correspondingly, the subduction of oceanic crust and sediment that had stewed in sea-water for millions of years provided a key to understanding why a different kind of rock, andesite and its intrusive analogue granite (or more precisely granodiorite), is found at subduction zones.

Plate tectonics accelerated a shift from geology being a largely descriptive science aimed mostly at the history of our planet to a quantitative physical science focused on the processes that have made the present-day Earth what it is.

Why so many young scientists?

As noted at the beginning in Chapter 1, plate tectonics was recognized when established giants were looking to the Moon. Of course, not everyone who brought us plate tectonics was young; Heirtzler, Hess, Oliver, and Wilson had established reputations. Yet, some were still students, like Atwater, Dickson, Francheteau, Herron, Pitman, and Vine, as were Irving when he did his initial palaeomagnetic work and Heezen when he and Marie Tharp were mapping the ocean floor in the 1950s; Plafker had not yet returned to graduate school. Many, like Isacks, Le Pichon, Matthews, McKenzie, Morgan, Parker, Sclater, and Sykes, had obtained their PhD degrees only a few years before they carried out their landmark studies, and the same would apply to Menard, when he recognized fracture zones in the 1950s, and to Cox, Doell, Dalrymple, MacDougall, Chamalaun, and Tarling when they did their work on reversals of the magnetic field. One might ask, how did the older folk miss the discoveries?

Many still wonder if Maurice Ewing, who more than anyone had gathered the data that established plate tectonics, ever accepted the idea. In the late 1960s Walter Pitman and Jeff Fox, one of the last students to work with Heezen at Lamont Geological Observatory, had planned a cruise to the Central Atlantic to study in some detail the intersection of a segment of ridge crest with a fracture zone. Ewing was delighted. At Ewing's house, just before Jeff was due to leave, Ewing's wife asked her husband what was the importance of this cruise, and he blurted out, 'Harriet! I just know that Walter and

Jeff are going to make a great discovery and will recover Paleozoic rocks and put this god damn plate tectonics nonsense to rest.'

Ewing died in 1974, and a few years later Frank Press, before becoming Jimmy Carter's Science Adviser, summarized Ewing's approach to science with a few tidbits of advice to young scientists. The first was: 'Make a better instrument or measure in a place where no one else has been, and a great discovery will come your way.' Ewing had obviously gathered marine geological data with that motivation in mind. Moreover, Ewing had urged Press, who had been one of the first, and most successful, students to work with Ewing at Lamont Geological Observatory, to design a new seismograph, one that could record longer periods of ground motion than most seismographs could. The Press–Ewing seismograph became the prototype for the World-Wide Standardized Seismograph Network, used by Sykes to confirm transform faulting and by countless others in refining plate tectonics.

Ewing's second piece of advice was, 'Do not hesitate to enter fields despite the giants who may be your competition, for it is the fresh analysis from a different vantage point that often leads to important new insights.' It is easy to imagine that younger people, whether oblivious to their ignorance or just eager to risk confronting a lion in its den, might blindly follow such advice without realizing it. It is worth remembering too that it was Wegener, a meteorologist, who entered a field in which he had little training and laid the groundwork for plate tectonics. Maybe if more of the older scientists had adhered to Ewing's suggestion, they too could have joined the fun.

Further reading

First, I urge readers to go to Tanya Atwater's website <http://emvc.geol.ucsb.edu/> and download animations that she prepared to illustrate many of the phenomena associated with plate tectonics. If a picture is worth a thousand words, her animations are packages of kilo-words.

Three books review aspects of plate tectonics well:

H. R. Frankel (2012), *The Continental Drift Controversy*, Cambridge University Press.

W. Glen (1982), *The Road to Jaramillo, Critical Years of the Revolution in Earth Sciences*, Stanford University Press, Stanford, CA, 459 pp.

H. W. Menard (1986), *The Ocean of Truth, A Personal History of Global Tectonics*, Princeton University Press, Princeton, 353 pp.

I also recommend the following technical papers:

T. Atwater (1970), Implications of plate tectonics for the Cenozoic evolution of western North America, *Geol. Soc. Amer. Bull.*, 81, 3513–36.

H. H. Hess (1962), History of the ocean basins, in *Petrologic Studies: A volume in honor of A. F. Buddington*, ed. A. E. J. Engel, H. L. James, and B. F. Leonard, Geol. Soc. Amer., 599–620.

B. Isacks, J. Oliver, and L. R. Sykes (1968), Seismology and the new global tectonics, *J. Geophys. Res.*, 73, 5855–99.

D. McKenzie, and R. L. Parker (1967), The North Pacific: an example of tectonics on a sphere, *Nature*, 216, 1276–80.

W. J. Morgan (1968), Rises, trenches, great faults, and crustal blocks, *J. Geophys. Res.*, 73, 1959–82.

F. J. Vine (1966), Spreading of the ocean floor: new evidence, *Science*, 154, 1405–15.

A. Wegener (1912), The origins of continents, *Geol. Rundsch.*, 3, 276–92, Translated by Roland von Huene (2002), *Int. J. Earth Sci. (Geol. Rundsch.)*, 91, S4–S17.

J. T. Wilson (1965), A new class of faults and their bearing on continental drift, *Nature*, 207, 343–7.

Index

C

D

E

F

G

H

I

J

K

L

M

N

O

P

Q

R

S

T

U

V

W

Z

76. Empire
77. Fascism
78. Terrorism
79. Plato
80. Ethics
81. Emotion
82. Northern Ireland
83. Art Theory
84. Locke
85. Modern Ireland
86. Globalization
87. The Cold War
88. The History of Astronomy
89. Schizophrenia
90. The Earth
91. Engels
92. British Politics
93. Linguistics
94. The Celts
95. Ideology
96. Prehistory
97. Political Philosophy
98. Postcolonialism
99. Atheism
100. Evolution
101. Molecules
102. Art History
103. Presocratic Philosophy
104. The Elements
105. Dada and Surrealism
106. Egyptian Myth
107. Christian Art
108. Capitalism
109. Particle Physics
110. Free Will
111. Myth
112. Ancient Egypt
113. Hieroglyphs
114. Medical Ethics
115. Kafka
116. Anarchism
117. Ancient Warfare
118. Global Warming
119. Christianity
120. Modern Art
121. Consciousness
122. Foucault
123. The Spanish Civil War
124. The Marquis de Sade
125. Habermas
126. Socialism
127. Dreaming
128. Dinosaurs
129. Renaissance Art
130. Buddhist Ethics
131. Tragedy
132. Sikhism
133. The History of Time
134. Nationalism
135. The World Trade Organization
136. Design
137. The Vikings
138. Fossils
139. Journalism
140. The Crusades
141. Feminism
142. Human Evolution
143. The Dead Sea Scrolls
144. The Brain
145. Global Catastrophes
146. Contemporary Art
147. Philosophy of Law
148. The Renaissance
149. Anglicanism
150. The Roman Empire
151. Photography
152. Psychiatry
153. Existentialism
154. The First World War
155. Fundamentalism
156. Economics
157. International Migration
158. Newton
159. Chaos
160. African History

161. Racism
162. Kabbalah
163. Human Rights
164. International Relations
165. The American Presidency
166. The Great Depression and The New Deal
167. Classical Mythology
168. The New Testament as Literature
169. American Political Parties and Elections
170. Bestsellers
171. Geopolitics
172. Antisemitism
173. Game Theory
174. HIV/AIDS
175. Documentary Film
176. Modern China
177. The Quakers
178. German Literature
179. Nuclear Weapons
180. Law
181. The Old Testament
182. Galaxies
183. Mormonism
184. Religion in America
185. Geography
186. The Meaning of Life
187. Sexuality
188. Nelson Mandela
189. Science and Religion
190. Relativity
191. The History of Medicine
192. Citizenship
193. The History of Life
194. Memory
195. Autism
196. Statistics
197. Scotland
198. Catholicism
199. The United Nations
200. Free Speech
201. The Apocryphal Gospels
202. Modern Japan
203. Lincoln
204. Superconductivity
205. Nothing
206. Biography
207. The Soviet Union
208. Writing and Script
209. Communism
210. Fashion
211. Forensic Science
212. Puritanism
213. The Reformation
214. Thomas Aquinas
215. Deserts
216. The Norman Conquest
217. Biblical Archaeology
218. The Reagan Revolution
219. The Book of Mormon
220. Islamic History
221. Privacy
222. Neoliberalism
223. Progressivism
224. Epidemiology
225. Information
226. The Laws of Thermodynamics
227. Innovation
228. Witchcraft
229. The New Testament
230. French Literature
231. Film Music
232. Druids
233. German Philosophy
234. Advertising
235. Forensic Psychology
236. Modernism
237. Leadership
238. Christian Ethics
239. Tocqueville
240. Landscapes and Geomorphology
241. Spanish Literature

242. Diplomacy
243. North American Indians
244. The U.S. Congress
245. Romanticism
246. Utopianism
247. The Blues
248. Keynes
249. English Literature
250. Agnosticism
251. Aristocracy
252. Martin Luther
253. Michael Faraday
254. Planets
255. Pentecostalism
256. Humanism
257. Folk Music
258. Late Antiquity
259. Genius
260. Numbers
261. Muhammad
262. Beauty
263. Critical Theory
264. Organizations
265. Early Music
266. The Scientific Revolution
267. Cancer
268. Nuclear Power
269. Paganism
270. Risk
271. Science Fiction
272. Herodotus
273. Conscience
274. American Immigration
275. Jesus
276. Viruses
277. Protestantism
278. Derrida
279. Madness
280. Developmental Biology
281. Dictionaries
282. Global Economic History
283. Multiculturalism
284. Environmental Economics
285. The Cell
286. Ancient Greece
287. Angels
288. Children's Literature
289. The Periodic Table
290. Modern France
291. Reality
292. The Computer
293. The Animal Kingdom
294. Colonial Latin American Literature
295. Sleep
296. The Aztecs
297. The Cultural Revolution
298. Modern Latin American Literature
299. Magic
300. Film
301. The Conquistadors
302. Chinese Literature
303. Stem Cells
304. Italian Literature
305. The History of Mathematics
306. The U.S. Supreme Court
307. Plague
308. Russian History
309. Engineering
310. Probability
311. Rivers
312. Plants
313. Anaesthesia
314. The Mongols
315. The Devil
316. Objectivity
317. Magnetism
318. Anxiety
319. Australia
320. Languages
321. Magna Carta
322. Stars
323. The Antarctic
324. Radioactivity
325. Trust

326. Metaphysics
327. The Roman Republic
328. Borders
329. The Gothic
330. Robotics
331. Civil Engineering
332. The Orchestra
333. Governance
334. American History
335. Networks
336. Spirituality
337. Work
338. Martyrdom
339. Colonial America
340. Rastafari
341. Comedy
342. The Avant Garde
343. Thought
344. The Napoleonic Wars
345. Medical Law
346. Rhetoric
347. Education
348. Mao
349. The British Constitution
350. American Politics
351. The Silk Road
352. Bacteria
353. Symmetry
354. Marine Biology
355. The British Empire
356. The Trojan War
357. Malthus
358. Climate
359. The Palestinian-Israeli Conflict
360. Happiness
361. Diaspora
362. Contemporary Fiction
363. Modern War
364. The Beats
365. Sociolinguistics
366. Food
367. Fractals
368. Management
369. International Security
370. Astrobiology
371. Causation
372. Entrepreneurship
373. Tibetan Buddhism
374. The Ancient Near East
375. American Legal History
376. Ethnomusicology
377. African Religions
378. Humour
379. Family Law
380. The Ice Age
381. Revolutions
382. Classical Literature
383. Accounting
384. Teeth
385. Physical Chemistry
386. Microeconomics
387. Landscape Architecture
388. The Eye
389. The Etruscans
390. Nutrition
391. Coral Reefs
392. Complexity
393. Alexander the Great
394. Hormones
395. Confucianism
396. American Slavery
397. African American Religion
398. God
399. Genes
400. Knowledge
401. Structural Engineering
402. Theatre
403. Ancient Egyptian Art and Architecture
404. The Middle Ages
405. Materials
406. Minerals
407. Peace
408. Iran
409. World War II

410. Child Psychology
411. Sport
412. Exploration
413. Microbiology
414. Corporate Social Responsibility
415. Love
416. Psychotherapy
417. Chemistry
418. Human Anatomy
419. The American West
420. American Political History
421. Ritual
422. American Women's History
423. Dante
424. Ancient Assyria
425. Plate Tectonics